Computational Probability
and Simulation

APPLIED MATHEMATICS AND COMPUTATION

A Series of Graduate Textbooks, Monographs, Reference Works

Series Editor: ROBERT KALABA, University of Southern California

Other Numbers in preparation

Computational Probability and Simulation

SIDNEY J. YAKOWITZ

The University of Arizona
Tucson, Arizona

▲▼ 1977

Addison-Wesley Publishing Company
Advanced Book Program
Reading, Massachusetts

London · Amsterdam · Don Mills, Ontario · Sydney · Tokyo

Library of Congress Cataloging in Publication Data

Yakowitz, Sidney J   1937-
  Computational probability and simulation.

  (Applied mathematics and computation ; no. 12)
  Includes bibliographical references and index.
    1. Probabilities--Data processing. 2. Digital com-
puter simulation. 3. Monte Carlo method. I. Title.
QA273.Y29   519.2   77-3002

ISBN 0-201-08892-4
ISBN 0-201-08893-2 (pbk)

Reproduced by Addison-Wesley Publishing Company, Inc., Advanced Book Program, Reading, Massachusetts, from camera-ready copy prepared by the author.

American Mathematical Society (MOS) Subject Classification Scheme (1970): 65C05, 68A55, 90C15, 90D35

Copyright © 1977 by Addison-Wesley Publishing Company, Inc.
Published simultaneously in Canada.

Manufactured in the United States of America

ABCDEFGHIJ-MA-7987

To my parents

CONTENTS

SERIES EDITOR'S FOREWORD

Execution times of modern digital computers are
measured in nanoseconds. They can solve hundreds of simultaneous ordinary differential equations with speed and
accuracy. But what does this immense capability imply with
regard to solving the scientific, engineering, economic, and
social problems confronting mankind? Clearly, much effort
has to be expended in finding answers to that question.

In some fields, it is not yet possible to write
mathematical equations which accurately describe processes
of interest. Here, the computer may be used simply to
simulate a process and, perhaps, to observe the efficacy of
different control processes. In others, a mathematical
description may be available, but the equations are
frequently difficult to solve numerically. In such cases,
the difficulties may be faced squarely and possibly overcome;
alternatively, formulations may be sought which are more
compatible with the inherent capabilities of computers.
Mathematics itself nourishes and is nourished by such
developments.

Each order of magnitude increase in speed and memory
size of computers requires a reexamination of computational
techniques and an assessment of the new problems which may
be brought within the realm of solution. Volumes in this

series will provide indications of current thinking
regarding problem formulations, mathematical analysis, and
computational treatment.

Probabilistic simulation is the activity of artifically
creating probabilistic processes for the purpose of solving
advanced problems in numerical analysis and operations
research. In contrast to numerical analysis itself,
simulation is strictly a product of the computer age, its
earliest developments being made by the pioneers of digital
computation. Since its inception, simulation has been a
standard tool of the trade in operations research, and its
value in numerical analysis remains firm as the mathematics
of simulation develops and computer capacity increases.

It is hoped that the present text, which seeks to
introduce students and nonspecialists to simulation
techniques while strengthening their grasp of probability
subjects, will help spur a generation of scientists oriented
toward the crucial subtleties of these powerful methods.

ROBERT KALABA

PREFACE

The intention of this textbook is to share with the
reader an experimental approach to studying the facts and
models of probability theory while introducing him to principles of probabilistic simulation.

This book is an outgrowth of a project begun over a
decade ago as a laboratory manual to accompany a standard
junior-senior course in probability theory. Through computer
experimentation, students were expected to illustrate the
major laws which were presented and proved in the class-
room sessions. While this volume differs from the laboratory
project in that much more sophisticated material on probabilistic simulation technology has been appended, I believe
that it remains a valuable complement or supplement to a
course in probability. Additionally, Computational Probability and Simulation will be useful to persons who, while
satisfied with their knowledge of probability theory, wish
to acquaint themselves with simulation technology. We have
offered as modern and inclusive a development of these subjects as is currently available, and our references will
lead readers to the research literature of the different
simulation subjects discussed.

Let us discuss the motivation of this text as an aid to
understanding probability theory. Through the use of random
number generator routines of digital computers, one has the
facility of performing a very large number of almost any
probabilistic experiment in a matter of seconds. The peda-
gogic intention of Computational Probability and Simulation
is to make use of this feature to study, in the outcomes of
a large number of experiments, trends which we are led to

expect by the fundamental laws of probability theory (e.g., laws of large numbers, limit theorems, properties of random walks).

Gnedenko and Kolmogorov, two giants of probability theory, write (1954, p. 1):

> In fact, all epistemologic value of the theory of probability is based on this: that large-scale random phenomena in their collective action create strict non-random regularity. The very concept of mathematical probability would be fruitless if it did not find its realization in the frequency of occurence of events under large-scale repetitions of uniform conditions (a realization which is always approximate and not wholly reliable, but that becomes, in principle, arbitrarily precise and reliable as the number of repetitions increases).

As the above excerpt implies, probability theory, like classical physics, makes strict assertions about measurable physical phenomena, but, in contrast to deterministic physics, a few measurements do not suffice to verify an assertion. Verification can come only through "large-scale repetition of the experiment under uniform conditions." In the past, such "large-scale repetition" was prohibitively difficult because typically it takes tens of thousands of repetitions before predicted trends are convincingly established. Modern computers have the capability of performing experiments at rates in the order of 100,000 per second. The opportunity that this capacity offers for observing probability theory in action is among the most interesting and worthwhile products of the computer age.

I have found that students (especially the gifted ones) are perplexed by probability theory when presented as a mathematical theory with a scattering of textbook-type

applications. It is especially difficult to explain to the
novice the meaning and significance of the limit theorems,
which, quite properly, are regarded by professional proba-
bilists as the focus of their theory. The proofs of the
limit theorems are tedious and give little insight into the
physical motivation of the theorems. Students seem to have
great difficulty recognizing how physical observables should
be associated with the variables of the limit theorems. In
Computational Probability and Simulation, we actually set
up experiments appropriate for the major probability models
and perform these experiments repeatedly (via digital computer techniques). For example, in investigating the central limit theorem, we will see how well the normal law
approximates observations of standardized sums of independent variates.

    In concert with and in addition to the aim of illus-
trating facts and models of probability theory, this text
provides a sound introduction to the subject of probabilistic
simulation. In this context, we will establish links between probabilistic and deterministic processes and use
these links to construct Monte Carlo methods for major numerical analysis problems (especially multivariate integration
and solution of linear equations) through our capacity to
rapidly generate random outcomes. Moreover, attention will
be paid to an important and popular application of the
methodologies of this book, namely simulation, or the experimental solution of probabilistic operations research and
optimization problems which are too difficult for direct
analysis. Simulation applications are illustrated by solution of decision problems and selection of strategies for
inventory control.

Our discussions of simulation topics will contain some
relatively sophisticated developments. Attention is paid
to modern methods which are less intuitive but more computationally efficient than early techniques. Readers seeking
only to round off their study of introductory probability
may wish to skip over some of this material. Specifically,
the sections which are primarily on aspects of simulation
are 1.2, 2.3, 2.4, 3.3, 4.4, 4.5, 5.2, 5.3, 6.4, and 6.7.
On the other hand 3.1, 3.2, 4.1, 4.2, 4.3, and 5.1 are
slanted much more toward the probabilistic subjects than
simulation methodology and may be bypassed by readers wishing
to concentrate on simulation theory.

Let us briefly survey the contents of this volume.
Computational Probability and Simulation begins with a careful discussion of the theory and limitations of pseudo-
random number generation, giving an algorithm for performing
this task. On the basis of the random-number generating
technique just presented, Chapter 2 gives an algorithm (with
analytic justification) so that, given any probability distribution function, one can generate independent observations
of the associated random variable. Specialized methods are
related for efficient generation of observations of the
common random variables.

Simple, intuition-defying gambling and random-walk
models studied by the noted probabilist William Feller (1968)
motivate Chapter 3. The random walk process is modified by
introducing a stopping variable, and by a device due to von
Neumann and Ulam, this process is used to solve simultaneous
linear algebraic equations.

Continuing to be motivated by gaming (a framework
wherein interesting problems of probability and operations

research can be stated precisely and plainly), we study
(Chapter 4) gambler's ruin problems, noticing through theory
and experimentation how rapidly the situation deteriorates
for the gambler as the game becomes subfair. The gambler's
ruin methodology is applied to prototype problems in actuary
science and inventory control.

In Chapter 5, we see how operations on Bernoulli sequences (which were fundamental to gambling and random walk
processes in Chapters 3 and 4) lead, asymptotically, to the
central limit theorem and to Brownian motion. Computer
simulation, coupled with elementary statistical analysis,
gives us some insight into how good these limiting approximations are for various sample sizes. The subject of
Gaussian random variates having been raised by the central
limit theorem, we go on to present efficient generation
techniques for Gaussian random vectors and time series.

In the final chapter (Chapter 6), we give a careful
theoretical study of the similarity of probability theory
and integration theory. From our analysis we are able to
establish a probabilistic interpretation of the integral
which allows us to associate a stochastic experiment with
a given integrand. Now we are able to approximate an integral (generally with useful and interesting probabilistic
bounds on error) by simulating the corresponding probability
experiment and analyzing a sequence of outcomes. Some theoretical and computational attention is given to the relative
merits of Monte Carlo integration (and its progeny, the
"number-theoretic method") in comparison to classical
numerical methods. The final part of this chapter outlines
the Monte Carlo approach to numerical solution of differential equations.

   As mentioned earlier, Computational Probability and
Simulation is intended to serve as a companion text for a
first course in probability theory. It is particularly
suited for use with the probabilistic portion of Lindgren,
Statistical Theory (1968), which is frequently cited for
definitions and results. Among other appropriate and com-
mendable texts are Birnbaum (1961), Feller (1968), Papoulis
(1965), Parzen (1960), Blum and Rasenblatt (1972), and Hogg
and Craig (1970). In light of the special and important
roles played throughout this book by Feller (1968) and
Lindgren (1968), we will refer to these works by simply
citing "Feller" and "Lindgren" followed by the page numbers
of interest. Each chapter contains a reference section,
but the publication information for Feller and Lindgren is
cited once and for all in the reference section of this
preface.
   The student who has no programming experience will be
able to see how the measurements resulting from the simulation experiments we have made are consistent with probability
theory. A reader who can program will be able to perform
his own experiments (such as suggested by the exercises
in this text). As exemplified by the many FORTRAN programs
in this text, generally the programs for computer simulation are short, provided the programmer has a firm grasp of
the probability theory principles involved.
   Unquestionably, I would be chanting the dogma of probability theory exactly as I learned it, had it not been my
good fortune to receive my first academic appointment with a
young department (Systems and Industrial Engineering). We
have no history or tradition, so that innovation, appraisal,
and reappraisal are our modus operandi. This inspiring

environment is due to the enlightened guidance of our former
Chairman, Dr. A. Wayne Wymore and my colleagues.   Also I
am extremely grateful to the many students (especially
Mssrs. Charles Ramsey, James Shiffrin, Robert Thuman, and
Sergio Davalos) who have helped in the formidable programming effort invested in the project on which this book
is based.

My special thanks go to Dr. John E. Krimmel, who provided the final versions of the programs and assisted in the
proofreading, and to Dr. John Halton, who suggested some
technical modifications.   Finally, Mrs. Nancy Greeson deserves
great credit for her patience and precision in typing the
many "final" versions of this book.

Sidney Yakowitz

REFERENCES FOR THE PREFACE

Birnbaum, Z., (1962). Introduction to Probability and Mathematical Statistics. Harper, New York.

Blum, E., and J. Rosenblatt, (1972). Probability and
   Statistics. Saunders, Philadelphia.

Feller, W., (1968). An Introduction to Probability Theory
   and Its Applications. Vol. I, 3rd ed. Wiley, New York.

Gnedenko, B., and A. Kolmogorov, (1954). Limit Distributions
   for Sums of Independent Random Variables. Addison
   Wesley, Reading, Mass.

Hogg, R., and A. Craig, (1970). Introduction to Mathematical Statistics. Macmillan, New York.

Lindgren, B., (1968). Statistical Theory. 2nd ed. Macmillan, New York.

Papoulis, A., (1965). Probability, Random Variables, and
   Stochastic Processes. McGraw-Hill, New York.

Parzen, E., (1960). Modern Probability Theory and Its
   Applications. Wiley, New York.

Computational Probability
and Simulation

CHAPTER 1

RANDOM PROCESSES AND RANDOM NUMBER GENERATORS

Tiger got to hunt,
Bird got to fly;
Man got to sit and wonder, "Why, why, why?"
Tiger got to sleep,
Bird got to land;
Man got to tell himself he understand.

Cat's Cradle
Vonnegut

1.0   BACKGROUND

A <u>random phenomenon</u> is characterized by the situation
that no matter how closely an experimenter duplicates the
conditions of his experiment, he is unable to predict, with
complete confidence, what values certain measurables of his
experiment will assume. The archetype of random phenomena

is the coin-tossing experiment. Classical physics (which includes Newton's laws of motion and optics, Gauss' and Maxwell's laws of electricity) as well as relativity theory restrict attention to deterministic phenomena wherein it is assumed that if the conditions of a sequence of identical experiments are not varied, the associated sequence of measurements too will not vary. Experiments not possessing this reproducibility property are termed random phenomena. Radioactive decay is a naturally occurring phenomenon which, in our present stage of understanding, demands classification as a random phenomenon. With deterministic phenomena, the physicist strives to find a law (function) $y = Q(x)$ which will state with accuracy that the measurement will be y whenever the conditions of the experiment are x. The attainment of the function Q represents the limit of understanding that one can have about the behavior of a phenomenon. (It is typical that the function is presented implicitly, being the solution to a given differential equation, as in Newton's and Maxwell's laws.) With random phenomena, such a description $Q(x)$ cannot be attained. The limit of behavioral knowledge is a probabilistic description of the measurement if the experiment with conditions x is performed. Thus with random phenomena, the role of $Q(x)$ is played by a probability function $P^x(A)$. The interpretation is that if the experiment is to be performed and the initial conditions are x, then the probability that the measured quantity y will be a member of the set A is $P^x(A)$. For further discussion on the distinction between deterministic and random phenomena, the reader may wish to refer to Chapter XIII of the book (Cramér, 1946).

With the above distinction between deterministic and
random processes, it is evident that, barring malfunction,
any general-purpose scientific computer is a deterministic
process, and no measurement such as printout should vary
unless changes are made in the program (the conditions of
the experiment). However, it is nevertheless possible for
a computer to print out a sequence of numbers {U(i)} which,
if you do not happen to know the algorithm whereby they
were calculated, appear to the casual observer as well as
to standard statistical tests to be a sequence of independent observations uniformly distributed* on the unit interval.
Such numbers U(i) are termed pseudo-random numbers, and they
will be the starting point for all of our numerical studies.

With digital computers, some discrete approximation
has to replace continuous random variables such as the uniform distribution. Thus we content ourselves with the goal
of generating the first N terms of a decimal expansion of
the uniform variable. If D(j) denotes the jth decimal of
the decimal expansion of the outcome of the uniform probability experiment, the event "D(j) = k" has probability
1/10 for each k = 0,1,2,...,9 and j = 1,2,3,.... This is
proven in Appendix 1.B. (In fact, the uniform distribution
may be characterized by this probability of the events
"D(j) = k" with the added stipulation that these events be
(stochastically) independent in j).

The rationale of pseudo-random number algorithms is
that successive numbers of common length N are constructed
in such a fashion that in the long run, each digit 0,1,...,9

---

*See Appendix 1.A for a brief description of the uni-
form probability distribution.

is expected to occur about 1/10 of the time in each decimal
place, and furthermore, the occurrence of a particular
digit at a particular decimal place is not expected to be
related to the digits occurring at the other decimal places
or with previously generated pseudo-random numbers. Without further information on the subject, quite likely most
people would anticipate that the last four digits of a
sequence of phone numbers, as well as successive blocks of
N digits in a decimal expansion of π, are related in this
fashion.

There is no proof that pseudo-random number generators
accomplish the objective of providing N-place decimal
expansions of uniformly distributed observations. In fact,
for reasons to be related at the close of this chapter,
there appears to be a theorem which implies that the program
for any sequence of uniformly distributed variables must be
essentially as long as the sequence itself. In short, all
feasible pseudo-random number generators are necessarily
faulty. Nevertheless, there are pseudo-random generators
that pass standard statistical tests pretty well and are
adequate and useful for many engineering applications. But
it is essential that the user should be aware of potential
shortcomings of any algorithm he employs. Generators should
constantly be scrutinized to assure that they suffice for
the uses to which they are being put.

1.1 RANDOM NUMBER ALGORITHMS

The method to follow is one of the earliest and
simplest random number generators.

ALGORITHM 1.1 MIDDLE-SQUARE RANDOM NUMBER GENERATOR.

Input.   N--the number of digits in the random number,
         M--an N-digit positive integer (known as the seed).

    0) Set J = 1
    1) X = M^2

    2) M = $\begin{cases} \text{middle N digits of X if "(number of digits} \\ \text{in X) - N" is even} \\ \text{middle N digits of 10X, otherwise} \end{cases}$

    3) U(j) = 0.M
    4) J = J + 1
    5) Go to 1.

Output. {U(j)}, a sequence of random numbers.

EXAMPLE: A COMPUTATION USING THE MIDDLE-SQUARE PSEUDO-
RANDOM NUMBER GENERATOR.

Input.   N = 3, M = 123

                 M = 123
                 X = 15129
                 M = 512
              U(1) = 0.512
                 X = 262144
                 M = 214
              U(2) = 0.214
                 X = 45796

                   . . . . . .

The middle-square random number generator was proposed and used by J. von Neumann (1951), who played a central role in the early development of computer simulation of probabilistic processes, as well as in programmable computer development in general. von Neumann is also known to engineers because of his celebrated contributions to game theory (such as the minimax theorem) and his studies in automata theory.

The middle-square technique can lead to highly non-random sequences. For example, if ever the first half of the digits are zeros, the sequence will soon after be for-ever composed of zeros. Whereas some scientists have re-ported obtaining satisfactory random sequences with the middle-square technique, it has been found to be inferior to other methods, principally because it tends to cycle early.

Let us discuss "cycling". The middle-square generator, like most random number generators, determines a sequence of random numbers $\{U(i)\}$ according to a rule of the form

$$U(n + 1) = f(U(n))$$

where the $U(i)$'s are numbers having some fixed number of digits. In the middle-square case, $f(\cdot)$ denotes the operation of squaring and extracting the middle digits. In view of the fact that there are only finitely many (say K) different values that the $U(j)$'s can assume, it is clear that eventually (in fact by $n = K$), there will be some j and $n > j$ such that $U(j) = U(n)$. And further $U(j + 1) = f(U(j)) = f(U(n)) = U(n + 1)$, and so forth. Thus $U(j + v) = U(n + v)$ for all positive integers v. The random numbers thereby

forever repeat themselves, the cycle U(j),U(j + 1),...,
U(n - 1),... occurring repeatedly ad infinitum. Random number generators which have cycles of short period (the period
being defined to be the minimum λ such that U(j) = U(j + λ))
are undesirable. For under these circumstances the same few
numbers occur over and over again and for that reason these
sequences do poorly in statistical tests for randomness as
well as in applications.

Early in the history of the Monte Carlo method, computer
scientists found experimentally that the middle-square technique generally led quickly to cycles of short period. By way
of illustration, in Table 1.1 the output is displayed of a
five decimal middle-square random number generator. The
program used for producing Table 1.1 had a routine in it
which located the onset of cycling as well as the beginning
of new periods. We have underlined these numbers. The input
M for this run was the leading digits in the decimal expansion of π, a popular "seed" for random number generators.

The scholarly opus Seminumerical Algorithms by D. Knuth
(1969) devotes 160 pages to technical aspects of random
number generation. In this work, a trick due to Floyd is
described which guarantees termination of a random number
sequence sometime during the first cycle. This is accomplished for any random number generator by computing at each
stage the quantities U(n) and U(2n), using stored values
U(n - 1) and U(2(n - 1)). The computation terminates when
U(2n) = U(n).

Table 1.1

MIDDLE-SQUARE RANDOM NUMBERS, SHOWING CYCLING

**CYCLING STARTS WITH THE   94TH NUMBER
AND THE PERIOD IS 33**

CYCLING OCCURS AGAIN WITH THE   127TH NUMBER

THE LIST OF RANDOM NUMBERS (READ ACROSS):

.8696	.6204	.4896	.9708	.2452	.0123	.5128	.2963
.7793	.7308	.4068	.5486	.0961	.3520	.3903	.2334
.4475	.0256	.5535	.6362	.4750	.5624	.6293	.6018
.2163	.6785	.0362	.1043	.0878	.0883	.9688	.8573
.4963	.6313	.8539	.9145	.6310	.8160	.5855	.2810
.8960	.2815	.9242	.4145	.1810	.2760	.6175	.1306
.7056	.7871	.9526	.7446	.4429	.6160	.9455	.3970
.7608	.8816	.7218	.0995	.0024	.7599	.7448	.4727
.3445	.8680	.3423	.7169	.3945	.5630	.6968	.5530
.5808	.7328	.6995	.9300	.4899	.0002	.9999	.9800
.0399	.9200	.6399	.9472	.7187	.6529	.6278	.4132
.0734	.8755	.6500	.2499	.2450	.0024	.7599	.7448
.4727	.3445	.8680	.3423	.7169	.3945	.5630	.6968
.5530	.5808	.7328	.6995	.9300	.4899	.0002	.9999
.9800	.0399	.9200	.6399	.9472	.7187	.6529	.6278
.4132	.0734	.8755	.6500	.2499	.2450	.0024	.7599
.7448	.4727	.3445	.8680	.3423	.7169	.3945	.5630
.6968	.5530	.5808	.7328	.6995	.9300	.4899	.0002
.9999	.9800	.0399	.9200	.6399	.9472	.7187	.6529
.6278	.4132	.0734	.8755	.6500	.2499	.2450	.0024
.7599	.7448	.4727	.3445	.8680	.3423	.7169	.3945
.5630	.6968	.5530	.5808	.7328	.6995	.9300	.4899
.0002	.9999	.9800	.0399	.9200	.6399	.9472	.7187
.6529	.6278	.4132	.0734	.8755	.6500	.2499	.2450
.0024	.7599	.7448	.4727	.3445	.8680	.3423	.7169

ALGORITHM 1.2 FLOYD'S ALGORITHM FOR A CYCLE-FREE FINITE
SEQUENCE OF RANDOM NUMBERS.

<u>Input.</u> U(1), and function f(·).

      0) Set n = 1, U(2n) = f(U(1))

      1) U(n + 1) = f(U(n))

      2) U(2(n + 1)) = f(f(U(2n)))

      3) If U(n + 1) = U(2(n + 1)), stop

      4) n = n + 1; go to 1

<u>Output.</u> $\{U(i)\}_{i=1}^{n}$

<u>Proof that termination occurs during the first cycle.</u> Let
M denote the time of onset of the first cycle and λ its
period. Thus

$$\{U(j)\} = U(1),\dots,U(M),\dots,U(M + \lambda - 1),U(M),\dots.$$

Define k to be the greatest integer such that $k\lambda < M$.
$h \equiv (k + 1)\lambda - M$. Then if N = M + h, we have

$$U(2N) = U(N)$$

since $2N - N = (k + 1)\lambda$ is a multiple of the period. Thus
termination occurs by time N and since $N = M + h < M + \lambda$,
termination must take place before the end of the first
cycle. On the other hand, since none of the values U(j),
j < M, recur, the stopping condition U(v) = U(2v) will not
be met for v < M.

Using developments in abstract algebra, useful random
number generators of the form U(n + 1) = f(U(n)) have been
devised which have the maximum possible period (determined
by the computer base raised to the word length). Furthermore,
the operations may be carried out quickly. The most popular
"maximum cycle length" generator is known as the linear
congruential random number generator and is determined by
the recursive rule:

$$X(n+1) = f(X(n)) = (aX(n) + c)(\text{modulo } K), \qquad (1.1a)$$

$$U(n+1) = \text{truncated decimal expansion of } X(n+1)/K. \quad (1.1b)$$

The meaning of (1.1a) is: multiply X(n) by a and add c;
then divide this result by K and set X(n + 1) equal to the
remainder. In (1.1a), all terms are integer; a and K are
assumed positive and c and X(n) are non-negative. (1.1b)
tells us that the nth random number is a decimal expansion
of X(n + 1)/K, which, of course, is between 0 and 1. Let us
state these operations formally:

ALGORITHM 1.3 LINEAR CONGRUENTIAL RANDOM NUMBER GENERATOR.

Input. X(0), a, c, and K (a, k > 0).

    0) Set J = 1, X = X(0)
    1) X = (aX + c) (modulo K)
    2) u(J) = truncated decimal expansion of X/K
    3) J = J + 1
    4) Go to 1

Output. {U(j)}.

In Program 1.1 (which is presented below), we generate
200 random numbers with a = 1001, c = 457, K = 1000, and
X(0) = 1000.

Program 1.1

Linear Congruential Generator

```
      PROGRAM LINCON (INPUT,OUTPUT,TAPE5=INPUT,TAPE6=OUTPUT)
C        LINCON GENERATES PSEUDO-RANDOM NUMBERS BY THE
C        LINEAR CONGRUENTIAL GENERATOR METHOD

C        X- THE STARTING VALUE (SEED)
C        A- THE MULTIPLIER
C        C- THE INCREMENT
C        K- THE MODULUS
C        U- THE SET OF GENERATED NUMBERS

      INTEGER X, A, C, K
      REAL U(200)

      READ(5,70) X, A, C, K
70    FORMAT(4I10)
      WRITE(6,80) X, A, C, K
80    FORMAT(1H1,14X," PSEUDO-RANDOM NUMBERS GENERATED BY"/
     1        15X," THE LINEAR CONGRUENTIAL METHOD"//
     2        19X," STARTING VALUE (SEED)    ",I10/
     3        19X," MULTIPLIER               ",I10/
     4        19X," INCREMENT                ",I10/
     5        19X," MODULUS                  ",I10///)

      DO 100 J=1,200
        X = MOD(A*X+C,K)
        U(J) = FLOAT(X)/FLOAT(K)
100   CONTINUE

      WRITE(6,81) U
81    FORMAT(10X,5F10.3)

      STOP
      END
```

PSEUDO-RANDOM NUMBERS GENERATED BY
THE LINEAR CONGRUENTIAL METHOD

STARTING VALUE (SEED)	1000
MULTIPLIER	1001
INCREMENT	457
MODULUS	1000

.457	.914	.371	.828	.285
.742	.199	.656	.113	.570
.027	.484	.941	.398	.855
.312	.769	.226	.683	.140
.597	.054	.511	.968	.425
.882	.339	.796	.253	.710
.167	.624	.081	.538	.995
.452	.909	.366	.823	.280
.737	.194	.651	.108	.565
.022	.479	.936	.393	.850
.307	.764	.221	.678	.135
.592	.049	.506	.963	.420
.877	.334	.791	.248	.705
.162	.619	.076	.533	.990
.447	.904	.361	.818	.275
.732	.189	.646	.103	.560
.017	.474	.931	.388	.845
.302	.759	.216	.673	.130
.587	.044	.501	.958	.415
.872	.329	.786	.243	.700
.157	.614	.071	.528	.985
.442	.899	.356	.813	.270
.727	.184	.641	.098	.555
.012	.469	.926	.383	.840
.297	.754	.211	.668	.125
.582	.039	.496	.953	.410
.867	.324	.781	.238	.695
.152	.609	.066	.523	.980
.437	.894	.351	.808	.265
.722	.179	.636	.093	.550
.007	.464	.921	.378	.835
.292	.749	.206	.663	.120
.577	.034	.491	.948	.405
.862	.319	.776	.233	.690
.147	.604	.061	.518	.975
.432	.889	.346	.803	.260
.717	.174	.631	.088	.545
.002	.459	.916	.373	.830
.287	.744	.201	.658	.115
.572	.029	.486	.943	.400

Care must be exercised in the choice of the parameters
a, c, and K in the linear congruential method in order to
obtain maximum-period sequences, to achieve computational
efficiency, and to obtain sequences that have the distribu-
tional properties of independent uniformly distributed
random variables. The following result of Hull and Dobell
(1962) gives a definitive solution to the choice of parameters for maximum cycle length. Clearly, the cycle length
cannot exceed the number of possible values of X (including
X = 0) in the algorithm. This bound is K, the modulus of
the congruential generator.

THEOREM 1: THE LINEAR CONGRUENTIAL SEQUENCE HAS PERIOD K
IF AND ONLY IF THE FOLLOWING HOLD:

1) c is relatively prime to K. (That is, they have no
   no common factors other than 1.)

2) a - 1 is a multiple of p, for every prime p dividing K.

3) a - 1 is a multiple of 4 if K is a multiple of 4.

The study by Marsaglia (1972) gives rules for finding
the period for any choices a, c, and K.

A great many generators in common use have zero as the
additive constant c in (1.1a). Such generators, which are
determined by the iterative form

$$X_{n+1} = aX_n (\text{modulo } K),$$
(1.1a')

are called multiplicative congruential generators. Since
all numbers factor zero, these generators fail to satisfy
the first condition of Theorem 1 and thereby fail to achieve
period K. The motivation of multiplicative generators is

that they are slightly faster than linear congruential generators by virtue of the elimination of the addition step. Their period, while not maximal, can be acceptably long. If λ is the period of a sequence determined by (1.1a') then

$$X_{n+\lambda} \equiv a^{\lambda} X_n \text{ (modulo K)}.$$

Therefore if λ is the period of the sequence, λ satisfies

$$a^{\lambda} \equiv 1 \text{(modulo K)}.$$

The modulus having been dictated (by computer word length, for example), an important objective to consider in choosing multiplicative congruential generator parameters a and X_0 is to attain as large a period as possible. Given K, any number a which gives the maximum (over all possible values of a) period is called, in the terminology of number theory, a <u>primitive element modulo K</u>. We will denote the associated maximum period value by $\lambda(K)$. Some important period values and relations include

$$\lambda(2^e) = 2^{e-2}, \text{ for } e > 3,$$

$$\lambda(p^e) = p^{e-1}(p - 1), \text{ for any prime number } p > 2.$$

If K has the prime factorization $K = \prod_{i=1}^{t} p_i^{e_i}$ then

$$\lambda(K) = \text{least common multiple of } (\lambda(P_1^{e_1}), \ldots, \lambda(P_t^{e_t})).$$

The modulus K having been selected, one obtains a
maximum period multiplicative generator (having period
$\lambda(K)$) by i) choosing X_0 relatively prime to K, and ii)
finding a multiplier a which is a primitive element modulo
K.   Knuth (1969, p. 19) describes convenient procedures for
finding primitive elements. Dieter, (1972) and references
therein provide supplementary information on choice and
performance of multiplicative and linear congruential
generators.

In the later chapters, we will use exclusively the CDC
random-number generator RANF, which is the multiplicative
congruential generator determined by

$$K = 2^{48},$$

$$U_0 = X_0/K = .170998384044023172,$$

and

$$a = 44485709377909.0.$$

From the preceding discussion, we may conclude that
this generator has period 2^{46}.

In addition to choosing a,c, and K according to
Theorem 1 and the preceding discussions, there are other
important considerations. In general, the division operation
in Step 1 of Algorithm 1.3 requires the most computer time
of the various operations. This division can often be avoided
altogether by choosing K wisely. For example, if K is the
word size W of the computer where

$$W = b^L, \quad (b = \text{computer base, } L = \text{word length})$$

then X = least L significant digits of aX. Further, but
less importantly, if a is of the form b^e, e > 1, then
multiplication can be achieved by shifting the digits of X
in Step 1. Knuth (1969, p. 11) discusses ways for avoiding
division when K = W + 1 or W - 1.

1.2 STATISTICAL TESTS OF RANDOM-NUMBER GENERATORS

A sequence {U(i)} of random numbers is in fact a non-
random sequence, and so at best, it can only approximate
the behavior of independent uniformly-distributed random
variables. In many cases, this approximation is very poor.
Computer scientist D. Knuth (1969, p. 4) writes, by way of
warning,

> Many random-number generators in use today
> are not very good. There is a tendency for people
> to avoid learning anything about the random-number
> generators; quite often we find that some old
> method which is comparatively unsatisfactory has
> blindly been passed down from one programmer to
> another, and today's users have no understanding
> of its limitations.

In the same vein, an issue of the Stanford University's
Academic Computing Services Bulletin (1975)* includes the
popular random-number generator RANDU among routines in the
IBM scientific Subroutine Package which are described as
"inaccurate, obsolete, and downright dangerous to use".
The RANDU generator is said to give "completely correlated
consecutive triples".

---

*See also the New Engineer, September 1975, p. 16.

Having established the need for critical testing of
random-number generators, it is unfortunately necessary to
point out that there is inadequate scientific basis for the
current testing procedures and only scant hope of ever finding definitive tests. The first fundamental roadblock on
the path toward random-number generator testing is that
present-day statistical theory is not equipped to test hypo-
theses containing nonrandom processes. In alluding to the
problem, Halton (1970, p. 29) writes

> One cannot apply statistical tests to
> deterministic sequences and expect the resulting
> probabilistic "levels of significance" to have
> more than a qualitative meaning. At best these
> tests give us crude empirical guides as to the
> general suitability of the tested sequences for
> Monte Carlo computations.

Secondly, even if one accepts that $\{U(n)\}$ can be treated
as a random sequence for purposes of statistical testing, we
are confronted with the inadequacy of present-day statistical
methodology for testing the hypothesis of independent, uniformly-distributed variables against the set of possible
alternatives. While certain tests have been used for this
statistical problem, these tests are not known to possess
those definitive properties (such as admissibility, or even
consistency) which inspire confidence among statisticians.
Establishing independence of a sequence is a very delicate
problem. For example, only recently (Dickey and Lienz (1970),
Yakowitz (1976), Denny and Yakowitz (1977)) have admissible
tests for the much simpler problem of establishing independence within the Markov hypothesis come to light.

Nevertheless, despite this gloomy backdrop, we will
proceed with standard statistical tests of random-number
generators. At least these tests are useful in discriminating

against very bad random-number generators. Furthermore,
in any given engineering application, only some of the
properties of independent, uniformly-distributed sequences
are really needed, and if the tests to follow do not
establish the requisite properties, often further tests can
be devised as needed.

Traditionally, in testing to see if {U(n)} is independently and uniformly distributed, uniformity and independence
are tested separately. We examine uniformity first, as the
existing theory for this test inspires more confidence than
the arguments supporting tests for independence.

Let $\{U(v)\}_{v=1}^{n}$ be a collection of numbers. The corresponding <u>empirical distribution function</u> F_n is defined by
the rule $nF_n(x)$ = Number of U(v)'s, $1 \leq v \leq n$, such that
$U(v) \leq x$. For any piecewise continuous function G, define

$$\| G \| \equiv \underset{0 < x \leq 1}{} |G(x)|.$$

The Glivinko-Cantelli theorem of probability theory (Loève
1955, p. 20) assures us that if the U(v)'s really are independent and uniformly distributed (i.e. have distribution
function F(x) = x, $0 \leq x \leq 1$) then with a probability of 1,

$$\| F_n(x) - F(x) \| \to 0.$$

In Figure 1.1, we have plotted $F_n(x)$ for sequences
associated with the CDC random-number generator RANF,
mentioned in the preceding section. One must conclude from
these plots that the random number sequence seems to be
converging to the uniform distribution function. Next we
ask whether the convergence is fast enough. The quantity

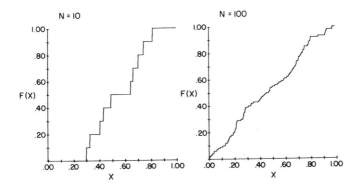

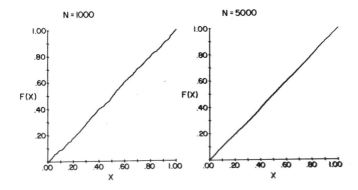

Figure 1.1   Empirical Distribution Functions from RANF

$||F_n(x) - x||$ is often abbreviated by D_n and is called the Kolmogorov-Smirnov statistic (or K-S statistic, for short). Under the hypothesis that the $U(j)$'s really are uniform and independent, its theoretical distribution function is well-known and frequently tabulated (e.g. Table VI in Lindgren). In Table 1.2 below, we have calculated D_n for various random number generators mentioned earlier. Of course, large values of D_n tend to cast doubt on the uniformity property. In addition to the test statistic, we have tabulated acceptance thresholds for tests at various levels. Thus, if the $U(j)$'s really are uniform, only 5% of the time will the D_n value exceed the threshold for level 0.05, which is denoted by $\tau_{.05}$ in Table 1.2.

If $F_1(x)$ distinct from the uniform distribution function x really describes the independent samples $U(i)$, then as n gets large, $D_n \rightarrow ||F_1(x) - x|| > 0$, and the tests will consistently reject the hypothesis of uniformity.

In Table 1.2, Generator 1 is linear congruential and has parameters $K = 10^5$, $a = 625$, $c = 457$, and $X(0) = 10^3$. It does not satisfy the conditions of maximum period. Generator 2 is linear congruential with parameters $a = 1001$, $K = 10^5$, $c = 457$, and $X(0) = 1000$. It does achieve maximum period (which is, of course, 10^5).

We have established, to some extent, the uniformity or "equidistribution" property of the CDC random-number generator. Next we turn our attention to the other distinguishing characteristic of random numbers: independence. This property is much more difficult to verify than uniformity, because of the rich class of processes which are alternatives to statistically independent sequences.

Table 1.2

K-S Test for Uniformity of Random Numbers

N = 100	Reject at α = α level if τ_α = $D_N > \tau_\alpha$	.20 .107	.15 .114	.10 .122	.05 .136	.01 .163

	Generator	RANF	GENERATOR 1	GENERATOR 2
	D_n	.073	.048	.129

N = 1000	Reject at α = α level if τ_α = $D_N > \tau_\alpha$	.20 .034	.15 .036	.10 .039	.05 .043	.01 .052

	Generator	RANF	GENERATOR 1	GENERATOR 2
	D_n	.023	.046	.009

N = 5000	Reject at α = α level if τ_α = $D_N > \tau_\alpha$	.20 .015	.15 .016	.10 .017	.05 .019	.01 .023

	Generator	RANF	GENERATOR 1	GENERATOR 2
	D_n	.011	.046	.003

In Figures 1.2 and 1.3, we have plotted pairs of consecutive random numbers from the CDC generator RANF. Under
independence, these points should be fairly evenly distributed over the unit square. The splotchy character of the
plot engenders wariness about the independence of the
samples.

A number of independence tests are cited in the Monte
Carlo literature (See Knuth (1969), Shreider (1966), or
Halton (1970)). None have the power and the appeal of the
Kolmogorov-Smirnov test for uniformity: there are no
theorems about these tests eventually distinguishing against

100 RANDOM POINTS

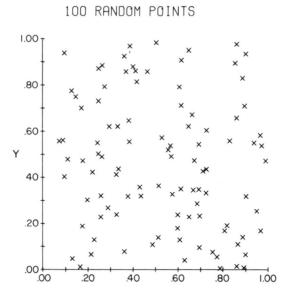

Figure 1.2   100 Pairs of Random Numbers from RANF

1000 RANDOM POINTS

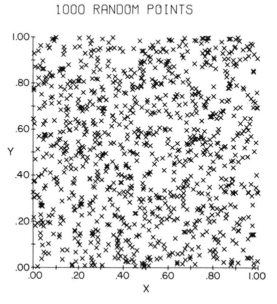

Figure 1.3   1000 Pairs of Random Numbers from RANF

any possible alternative. We apply one of the popular
tests, the run test, which has a distinctive history in
that many prominent statisticians have made contributions
to the distribution theory behind this test. Some of the
basic papers on the theory of run tests include Barton and
Mallows (1965) and Levene and Wolfowitz (1943).

There are several different definitions of a "run" in
vogue. The definition which was employed in our numerical
study is that a "run up" terminates at U_j if $U_{j+1} < U_j$. By
way of example, the runs up in the following sequence of
numbers are enclosed in parenthesis:

$$(21,33)(26,80,99)(60)(50,57)(52).$$

In this example, we have two runs up of length 1, two of
length 2, and one of length 3. Under independence, the
joint distribution of the number of runs up of varying
length has been calculated (and is known to be independent
of the distribution of the variable U(j) itself). Run tests
are tests which measure how well run counts in an observed
sequence compare with the theoretical distribution. In
Knuth (1969, p. 60) the specific run test statistic

$$V = 1/n \ \Sigma_i (R(i) - nb_i)(R(j) - nb_j)a_{ij}, \ 1 \le i, \ j \le 6$$

is employed. For $1 \le i \le 5$, R(i) denotes the number of runs
up of length i in the observed sequence of n numbers and
R(6) is the number of runs up of length at least 6. The
constants a_{ij} and b_i are given in Appendix 1-C. Asymptotically, V has the chi-square distribution with 6 degrees of
freedom, under the hypothesis that the U(i)'s are independent

and identically distributed. In Table 1.3 this test
statistic is applied to the three random-number generators
studied in Table 1.2. The higher the number V, the more
suspect the independence hypothesis becomes. In all cases
but the RANF generator, the performance is miserable. Even
the RANF generator fails at all levels higher than 1%.

Table 1.3

Run Test for Independence of Random Numbers (Based on 5000
Random Numbers)

Test at level α: Reject independence hypothesis if
  $v > \tau_\alpha$

Chi-Square thresholds:	α =	.20	.10	.05	.01
	τ_α =	8.56	10.6	12.6	16.8
	RANF	Generator 1[i]		Generator 2[ii]	
Test Statistic V:	16.78	354.7		31.04	

Congruential Generator Parameters (multiple, increment, modulus)
  i) Generator 1:   1001, 457, 10^5
 ii) Generator 2:   10001, 76711310, 10^{12}

1.3 SUMMARY

There is strong reason to believe that the outputs of
random number generators in operation today fall short of
having the properties of an independent sequence of uniform
variables. About linear congruential generators, Marsaglia
(1972) states flatly, "[they] are not suitable for precision

Monte Carlo uses." This work goes on to provide a sound
basis for this assertion. Perhaps the best medium for
giving substance to these doubts is to discuss the
Kolmogorov-Chaitin measures of randomness of finite sequences
of elements taken from a finite set W (such as the alphabet
of a computer). Our discussion will be unwholesomely sketchy.
For a supplementary introduction to this important subject,
the reader is advised to consult Chaitin's (1975) Scientific
American article.

Let W be a finite set. The randomness of a sequence
$(w_1, w_2, \ldots, w_n)$ of numbers in W is defined to be $R(\{w_i\})$
= length of shortest program of a "program-efficient"
universal Turing machine which produces $w_1, w_2, \ldots, w_n$.

Due to space constraints, it is not feasible for us to
define "universal Turing machine". Instead, we refer the
reader to Minsky's (1967) book for a good source of information on the needed automata theory and to Chaitin (1966,
1975) for a comprehensive development of the notion of
randomness discussed here. The above definition of randomness seems to be the most satisfying definition currently
available. For example, Cover (1974) proves that there is
no favorable (winning) betting scheme for placing bets at
even odds on consecutive terms in sequences with full "randomness." This definition can, with only slight loss in
precision, be rephrased in terms of more familiar concpts.
It is correct to say that with respect to some given computer
and some specified programming language, the randomness of a
sequence $(w_1, w_2, \ldots, w_n)$ is the length of the shortest program
which causes this sequence to be printed out. It has been
proven that sequences with the same degree of randomness as
independent, uniform (on W) distributions have the properties
one expects of random sequences.

Inasmuch as most random-number generators have pitifully short programs, the implication of the Kolmogorov-Chaitin theory is that their output sequences cannot achieve a very high degree of randomness. While this assertion may seem contradictory to the fact that some of the better random-number generators pass statistical tests, the interpretation of this author of the test performance of random-number generators is that this provides additional confirmation to his long-standing suspicion that tests for independence are not very discerning, except, perhaps, on extremely long samples.

But for our purposes of demonstrating central issues of probability theory and introducing the readers to the Monte Carlo method, the above discouraging developments concerning the ultimate randomness of the random-number generators is not an insurmountable obstacle. A conclusion of our studies is that the CDC random-number generator RANF is fairly successful in mimicking some of the salient properties of truly independent sequences. In the sequel of this book, we see that it serves us well: Our experimental results are not contrary to those which probability theory leads us to anticipate. Further, in many applications, only some of the properties of independent sequences are really needed.

APPENDIX 1.A

THE DEFINITION OF THE UNIFORM PROBABILITY LAW

In this experiment, we take the "uniform law" to mean the uniform distribution on the unit interval. Thus the sample space is [0, 1], the field of events is generated by subintervals [a, b] $\subset$ [0, 1], and the probability law is uniquely defined by the probability function P[[a, b]] = b - a, $0 \leq a \leq b \leq 1$. Observe that the cumulative distribution function of the uniform law is

$$F(x) = \begin{cases} P[[-\infty, x]] = P[[0, x]] = x - 0 = x, & 0 \leq x \leq 1, \\ 0, & x \leq 0, \\ 1, & x \geq 1. \end{cases}$$

The associated density function f(x) = dF(x)/dx is

$$f(x) = \begin{cases} 1, & x\epsilon \ [0, 1] \\ 0, & \text{otherwise.} \end{cases}$$

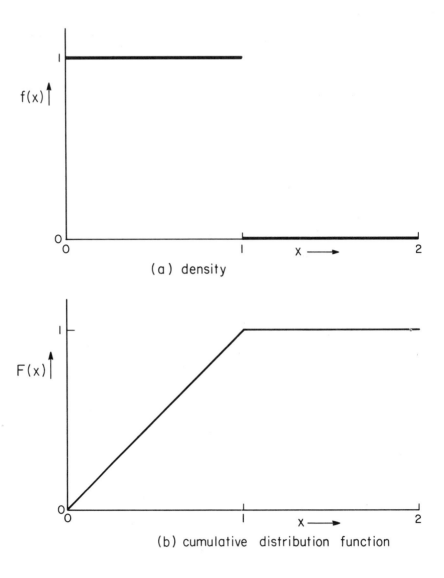

(a) density

(b) cumulative distribution function

Figure 1.4   Uniform Density and Distribution Functions

APPENDIX 1.B*

THE DISTRIBUTION OF THE DECIMALS

IN THE EXPANSION OF THE UNIFORM VARIABLE

THEOREM.   Let $D(j)$ denote the $\underline{jth}$ decimal in the decimal
expansion of the uniform variable U.   That is,

$$U = \sum_{j \geq 1} D(j)10^{-j}.$$

Then

    a) With probability 1, the sequence $\{D(j)\}$ is unique.

    b) $P[D(j) = k] = 1/10$, for $0 \leq k \leq 10$ and $j = 1,2,\ldots.$

    c) The collection $\{D(j): \ j \geq 1\}$ of (discrete)
        random variables is independent.

---

*This Appendix is somewhat more sophisticated than the
rest of this chapter.

Proof of (a). Non-unique decimal expansions have the form
$a_1 a_2 \ldots a_N 999 \ldots$ There are only countably many such numbers
(inasmuch as they are a subset of the rational numbers).
Let $X_1, X_2, \ldots$, be an enumeration of these numbers with non-
unique expansions. Then for any $\varepsilon > 0$,

$$P[U \in \{X_1, X_2, \ldots\}]$$

$$\leq P_U[X_1 - \varepsilon, X_1 + \varepsilon)] + P_U[(X_2 - \varepsilon^2, X_2 + \varepsilon^2)]$$

$$+ P_U[(X_3 - \varepsilon^3, X_3 + \varepsilon^3)] + \cdots < 2(\varepsilon + \varepsilon^2 + \varepsilon^3 + \cdots)$$

$$= 2\varepsilon/1 - \varepsilon. \tag{1}$$

Since inequality (1) holds for all $\varepsilon > 0$, the probability on
the left hand side is 0.

Note. This proof shows that for any continuous random variable, the probability of a countable set is 0.

Proof of (b). Let j' be the least integer such that for some
k, $P[D(j) = k] \neq 1/10$, and k' be the smallest such k associated with j'. Then

$$F_U(k'10^{-j'}) = 10^{-j'} \sum_{k=0}^{k'} P[D(j) = k]$$

$$= [\max \{(k' - 1), 0\} + P[D(j') = k']]10^{-j'}$$

$$\neq k'10^{-j'}$$

in contradiction to the definition of the cumulative distribution function of the uniform variable (Appendix 1.A).

<u>Proof of (c)</u>. The basic argument is similar to the proof
of (b). Let M be some integer such that the set $\{D(j): j \leq M\}$ of random variables are mutually stochastically dependent (see Lindgren, Section 2.1.10 for criteria for
independence). Thus, from the definition of "dependent,"
there must be some integer M-tuple, $\underline{k} = (k_1,\ldots,k_M)$ such
that

$$P[D(j) = k_j, j \leq M] \neq \prod_{j=1}^{M} P[D(j) = k_j] = 10^{-M}. \quad (2)$$

Let $\underline{a}$ be the M-tuple satisfying (2) that yields the least
value of $\Sigma\, k_j 10^{-j}$. $x \equiv \Sigma\, a_j 10^{-j}$. Say that $\underline{k} < \underline{a}$ if $k_j \leq a_j$,
with strict inequality for at least one value j. $S \equiv \{\underline{k}:$
$\underline{k} < \underline{a}\}$. The reader may verify that S has $10^M x - 1$ elements.
From the definition of $\underline{a}$,

$$P[D(j) = k_j, j \leq M] = 10^{-M}, \text{ all } \underline{k} \in S. \quad \text{Thus if } \underline{D}$$
$$= \{D(j)\}_{j=1}^{M}$$
$$F_U(X) = P[\underline{D} < \underline{a}] + P[\underline{D} = \underline{a}]$$
$$= 10^{-M}(10^M x - 1) + P[\underline{D} = \underline{a}]$$
$$= x - 10^{-M} + P[\underline{D} = \underline{a}],$$

which, in light of the fact that $P[\underline{D} = \underline{a}]$ satisfies inequality (2), yields the desired result that $F_U(x) \neq x$,
which contradicts the definition of the distribution of U.

APPENDIX 1.C

COEFFICIENTS FOR THE RUN TEST

$A = \{a_{ij}\}_{1 \le i, j \le 6}$

$$= \begin{bmatrix} 4529.4 & 9044.9 & 13568 & 18091 & 22615 & 27892 \\ 9044.9 & 18097 & 27139 & 36187 & 45234 & 55789 \\ 13568 & 27139 & 40721 & 54281 & 67852 & 83685 \\ 18091 & 36187 & 54281 & 72414 & 90470 & 111580 \\ 22615 & 45234 & 67852 & 90470 & 113262 & 139476 \\ 27892 & 55789 & 83685 & 111580 & 139476 & 172860 \end{bmatrix} .$$

$b = \{b_i\}_{i=1}^{6} = (\frac{1}{6}, \quad \frac{1}{24}, \quad \frac{11}{120}, \quad \frac{19}{720}, \quad \frac{29}{5040}, \quad \frac{1}{840})$

EXERCISES

1) Invent some "sensible" random-number generator whose
   output sequence {U(n)} satisifies a recursive relation
   of the form U(n + 1) = f(U(n)), but is nevertheless neither
   a linear congruential nor a middle-square generator.

2) Generate a sequence $\{U(n)\}_{n=1}^{N}$ of random numbers according to any rule of your choosing. For each U(n) and for
   some fixed integer K, define the integer-valued pseudo-
   random variable X(n) by

   X(n) = Integer Part of (KU(n)).

   Compare for various sequence lengths N, and for various
   numbers k, $0 \leq k \leq K$, the relative frequency of the
   event "X(n) = k" with its theoretical value, which of
   course is 1/K. Note: The relative frequency of occurence of an event E in a sequence of N observations

$\{X(i)\}_{i=1}^{N}$ is defined to be [Number of times $X(i) \in E$, $0 \leq i \leq N]/N$.

3)   (Continuation for readers who have some acquaintance
      with statistics.) Assume the $X(i)$'s in Problem 2 are
      independent observations of some variable X. Design
      and perform a chi-square test of the hypothesis that
      $P[X = k] = 1/K$, $0 \leq k \leq K$, using the integer sequence
      $\{X(i)\}_{i=1}^{N}$ of Problem 2 as the data for the test.

4)   Assume a given computer has a word length of L bits.
      How many distinct random number sequences $\{U(n)\}$ de-
      termined by a recursive rule of the form

$$X(n + 1) = f(X(n))$$
$$U(n) = X(n)/2^{L}$$

      can be programmed? Assume that $X(i)$'s are representable
      as computer words and $X(0)$ is fixed. Different se-
      quences are obtained by using different rules f.

5)   (Continuation.) How many of the random-number sequences
      in Problem 4 attain the maximum possible cycle period?

6)   (A random random-number generator.) The sequences in
      Problem 4 are characterized by f. Let F denote the set
      of possible functions f. Suppose a function f' is
      selected from F at random. That is, every function in
      F has the same probability of being selected as f'.
      Let λ denote the least integer such that $U(\lambda) = U(j)$ for
      some $j < \lambda$.

a) What is the probability that $\lambda < V$, for any
number V?

b) Develop an expression for the expectation of λ.
(This problem is adopted from Knuth (1969, p. 8).)

c) Compute the probabilities and expectation in a,
b for L = 24. (Hint: Use Stirling's formula
for the factorials.)

7) Think of some (preferably complicated) event in the
sample space for pairs (U, V) of independent uniform
(on [0, 1]) random variables. (For example, E might
be the event that (U, V) lies in the first quadrant
of the unit circle). Compute the probability of the
event E. Then choose some random number generator and
use it to construct a sequence $\{(U(n), V(n))\}_{n=1}^{N}$ of
pairs of random numbers; 2N random numbers should be
used in obtaining N pairs. Compute the relative frequency of occurence of E, and compare this with the
theoretical value you obtained earlier.

8) Perform the K-S test for uniformity and the run test
for independence (as described in Section 1.1) on the
random number generator of your computer. Compare
your test statistics with those in Tables 1.1 and 1.2.

9) Let L be a fixed positive integer representing the
binary word length of some computer. Let F denote the
set of functions f mapping $S \times S \rightarrow S$, where S is the set
of binary L-tuples.

a) How many functions are in F?

b) Specify one such function in F and use it to obtain
random numbers by using the recursive rule

$$X(n + 1) = f(X(n), X(n - 1)), \quad U(n) = X(n)/2^L.$$

(Choose $X(0)$ and $X(1)$ arbitrarily.)
With $U(n)$ determined as above, define

$$Z(n) = \text{Integer Part } (10U(n)).$$

Simulate a sequence of 10,000 random numbers and find the relative frequency that

$$Z(n) = j, \quad 0 \leq j < 10.$$

10) (Continuation.) How many functions in F above yield cycles of maximum period? What is this maximum period?

11) (Continuation.) Give a modified version of Floyd's method for stopping a generator of the form given in Problem 8. The halting should occur sometime during the first cycle, and the memory requirement should be fixed and not large.

REFERENCES FOR CHAPTER 1

Academic Computing Services Bulletin, (1975). Vol. III-8.
    Stanford University, California, pp. 10-11.

Barton D., and C. Mallows, (1965). "Some Aspects of the
    Random Sequence." Ann. Math. Statist., 36, pp. 236-
    260.

Chaitin, G., (1966). "On the Length of Programs for
    Computing Finite Binary Sequences." Journal for
    Automatic Computing Machines, 13, pp. 547-569.

Chaitin, G., (1975). "Randomness and Mathematical Proof."
    Scientific American, June, pp. 47-52.

Cover, T., (1974). "Universal Gambling Schemes and the
    Complexity Measures of Kolmogorov and Chaitin."
    Technical Report No. 12, Dept. of Statistics, Stanford
    University.

Cramér, H., (1946). Mathematical Methods of Statistics.
    Princeton University Press, Princeton, New Jersey.

Denny, J., and S. Yakowitz, (1977). "Admissible Run-
    Contingency Type Tests." J. American Statistical
    Assoc., to appear.

Dickey, J., and B. Lietz, (1970). "The Weighted Likeli-
    hood Ratio and Sharp Hypotheses About Chances, the
    Order of a Markov Chain." Ann. Math. Statist., 41,
    pp. 214-225.

Dieter, V., (1972). "Statistical Interdependence of
    Pseudo-Random Numbers Generated by the Linear
    Congruential Method," in Applications of Number
    Theory to Numerical Analysis, ed. S. Zaremba.
    Academic Press, New York, pp. 289-318.

Fortran Mathematical Library, (1973). Documentation,
    Control Data Corp., Minneapolis, Minn., pp. 193-196.

Halton, J., (1970). "A Retrospective and Prospective
    Survey of the Monte Carlo Method." Soc. Indus. Appl.
    Math. Rev., 12, pp. 1-63.

Hammersley, J., and P. Handscomb, (1964). Monte Carlo
    Methods. Methuen and Company, London.

Hull, T., and A. Dobell, (1962). "Random Number Generators."
    Soc. Indus. Appl. Math. Rev., 4, pp. 230-254.

Knuth, D., (1969). Seminumerical Algorithms. Addison
    Wesley, Reading, Massachusetts.

Lehmer, D. H., (1949). "Mathematical Methods in Large Scale
    Computing Units." Proc., Symp. on Large Scale Digital
    Calcul. Machinery, Harvard University Press, pp. 141-146.

Levene, H., and J. Wolfowitz, (1944). "The Covariance
    Matrix of Runs Up and Down." Ann. Math. Statist., 15,
    pp. 58-69.

Loève, M., (1955). Probability Theory. Van Nostrand,
    Princeton, New Jersey.

Marsaglia, G., (1972). "The Structure of Linear Congruential
    Sequences." In Applications of Number Theory to Numeri-
    cal Analysis, ed. S. Zaremba. Academic Press, New
    York, pp. 249-286.

Minsky, M., (1967). Computation. Prentice Hall, Englewood
    Cliffs, New Jersey.

Shrieder, Yu., Ed., (1966). The Monte Carlo Method.
    Pergamon Press, New York.

Von Veumann, J., (1951). "Various Techniques Used in Con
    nection with Random Digits." NBS Applied Math. Series,
    No. 12, pp. 36-38.

Yakowitz, S., (1976). "Small Sample Hypothesis Tests of
    Markov Order, with Application to Simulated and Hy-
    drologic Chains." J. American Statist. Assoc., 71,
    pp. 132-136.

Note. As stated in the preface, Feller and Lindgren are
referenced there, once and for all.

CHAPTER 2

SIMULATION OF PROBABILITY EXPERIMENTS

Above all, Sir, stand by me at my lectern;...
save me from slip of tongue and lapse of memory, from twice-told joke and unzippered fly.
Doctor of doctors, vouchsafe unto me examples
of the Unexampled, words to speak the Word-
less; be now and ever my visual aid, that upon
the empty slate of these young minds I may
inscribe, bold and squeaklessly, the Answers!

Giles Goat-Boy
John Barth

2.0   BACKGROUND

The object of Chapter 2 is to reveal how random-number
generators can be used to supply a sequence of observations
$\{x_i\}$ of any given random variable X.   Roughly speaking, if
the random numbers are really independent and uniform,
this computer-generated sequence is statistically

indistinguishable from a sequence obtained by observing repetitions of the corresponding actual random physical phenomenon. The great significance of the computer's capacity to artificially perform experiments is that it can perform a great many experiments in a very short time (typically tens of thousands in a second). One cannot hope to gather a comparable quantity of data in a reasonable time in any other fashion.

The methods of probabilistic simulation are dedicated to exploiting this ability of modern computers to generate vast amounts of probabilistic data. Through analysis of this data (computers are called upon to do this also), we will investigate trends which probability theory predicts we should be able to find and which serve as foundation for the subject of statistics. For eloquent statements on the significance of these trends, we remind the reader of the excerpt from Gnedenko and Kolmogorov recorded in our Preface, and also relate below a statement due to the probabilist and statistician H. Cramer [1946, Section 13.1].

> We have seen that, in a sequence of random experiments, it is not possible to predict individual results. These are subject to irregular random fluctuations which cannot be submitted to exact calculation. However, as soon as we turn our attention from the individual experiments to the whole sequence of experiments, the situation changes completely, and an extremely important phenomenon appears: In spite of the irregular behavior of individual results, the average of long sequences of random experiments show a striking regularity.

(Cramer's italics)

Our primary purpose in the remaining chapters of this book will be to examine and utilize this "striking regularity" of "long sequences of random experiments." First,

our algorithm for generating arbitrary random variables will
be revealed (with some simple illustrations of its use), and
then will follow the requisite theory (especially Appendix
2.1) to demonstrate that the algorithm "works." We shall
also give consideration to numerical analysis aspects of
the efficient production of random variables. Below, $\{u_i\}$
denotes a sequence of random numbers and $F(x) = P[X \leq x]$.
That is, F is the probability distribution function of the
variate X.

ALGORITHM 2.1 A UNIVERSAL RANDOM VARIABLE GENERATOR.

<u>Inputs</u>. $F(x)$, $\{u_i\}$.

    0) Set $i = 1$.

    1) For $u = u_i$, find x such that

$$x = \text{minimum} \{y: \ F(y) \geq u\}. \tag{2.1}$$

    2) Set $x_i = x$.

    3) $i = i + 1$. Go to 1.

<u>Output</u>. The sequence $\{x_i\}$.

    Appendix 2.A contains a proof that if the random num-
bers u_i are uniform*, the x_i's will have distribution F.

---

*In view of experimentation results in Chapter 1, we assume
here and hereafter that a sequence of random numbers is distributed as a sequence of independent observations of the
uniform variable on [0, 1]. Statistical tests in the sequel
serve indirectly as further check on the random-number
generator. The CDC routine RANF is used as a random-number
source during the remaining programs in this book.

It is instructive to study a graphical interpretation
of the algorithm. Consider an arbitrary cumulative distribution function F(x) illustrated in Figure 2.1 below.

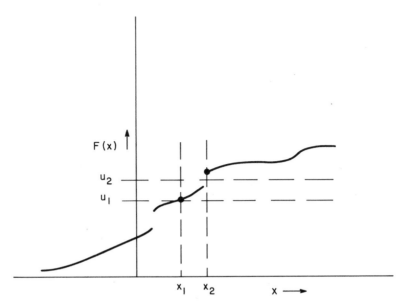

Figure 2.1   Illustration of Algorithm 2.1

Having generated a random number u, to get the associated
observation x of the random variable having the distribution function F(x), we trace a horizontal line passing
through (0, u). If this horizontal line intersects the
graph of F(x) (as does the line labeled u_1), then x_1, the
abscissa of the point of intersection, should be taken for
the value x. If on the other hand, the line passing through
(0, u) does not intersect F(x) (see, for example, the line
marked u_2), then F has a discontinuity such that all values
F(x) for x <u>less than</u> some number, say x_2, are less than u

and for x <u>greater than or equal to</u> x_2, $F(x)$ is greater than
u. In this case, x_2, the smallest value of x such that $F(x)$
is not less than u, is the number selected by the algorithm.
In theory, these two cases exhaust the possibilities, for
if the random number is truly uniformly distributed, the
probability is 0 that the horizontal line through (0, u)
will intersect a horizontal portion of $F(x)$. The reason
for this is that there can be only countably many disjoint
horizontal segments in the graph of a real function, and
thus the set of ordinates of these segments is countable.
Under the uniform law, any countable event has probability
0. (See the proof contained in Appendix 2.B for a demonstration of this fact.) In practice, however, because of
discretization of random numbers, in some cases u may intersect a horizontal portion of the graph of F. One will have
to devise a method for finding or approximating the left
endpoint of the interval, in such a happenstance.

    In equation (2.1) of step 1 of the algorithm, when we
write "$x = \min \{y: F(y) \geq u\}$" it is implied that $F(x) \geq u$.
(Otherwise, it would be necessary to replace "minimum" by
"greatest lower bound" since the set would not necessarily
contain its minimizing point.) The following argument
justifies the use of "minimum" in step 2. Let $\{y_i\}$ be a sequence of numbers such that y_i converges to x and for all
i, $y_i > x$. Since distribution functions are nondecreasing
functions (Lindgren, p. 67),

$$F(y_i) \geq u \text{ for all i.} \tag{2.2}$$

Also, distribution functions are continuous from the right
(Lindgren, p. 67). This implies $\lim_{i \to \infty} F(y_i) = F(x)$.

Thus

$$F(x) \geq u, \qquad\qquad\qquad (2.3)$$

since $F(x)$ is the limit of a sequence of numbers, none of which is less than u.

    Let us continue our examination of implementation of the algorithm by using it to simulate a few of the popular random variables.

2.1   SIMULATION OF DIE THROWING

    The standard probability model for throwing a fair die is the experiment whose sample space is $\{1,2,\ldots,6\}$ and whose probability law assigns the same probability (which must necessarily be 1/6) to each element of the sample space. The cumulative distribution function of the associated random variable X is graphed below.

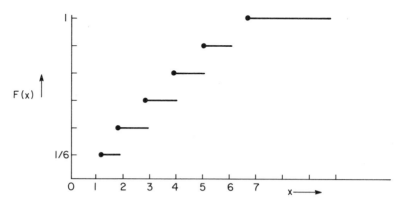

Figure 2.2   Distribution Function for a Die Throw

From Figure 2.2 it may be concluded that if the random number u is between 0 and 1/6, the value x given by Algorithm
2.1 is 1; if u is between 1/6 and 1/3, x = 2, etc. Program
2.1 first simulates a relatively small number of casts of
the die and prints out the outcome of each of these tosses.
Then a great many casts are simulated and the relative frequency of occurrence of the numbers 1,2,...,6 are tabulated
and compared to the probability of these numbers (which is
0.1666...).

    Let $\overline{X}_j(n)$ denote the relative frequency of j points
showing in n tosses, $1 \leq j \leq 6$. A common measure of dis-
crepancy between the relative frequency and the underlying
hypothesized probabilities p_j is given by the chi-square
statistic

$$\chi_n^2 = \sum_{i=1}^{6} n(\overline{X}_j(n) - p_j)^2/p_j.$$

If the p_j's are incorrect, $\overline{X}_j(n)$ will converge to the actual
probability of j, which is unequal to p_j, and χ_n^2, being
roughly proportional to n, will therefore increase without
bound as the number of tosses increases. Also, if distur-
bances are introduced by a faulty random-number generator
or an incorrect rule for simulating the die tosses, one
would intuitively anticipate that χ_n^2 would increase without
bound as n increases. On the other hand, if all the assumptions are fulfilled, it is known (Lindgren, p. 325f) that

χ_n^2 converges* with increasing n to a certain random variable, the χ^2 variable with five degrees of freedom. From widely-published tables (such as Lindgren, Table II), we have that $P[\chi^2 < 7.29] = 0.8$. We see in the print-out of Program 2.1 that most of the values of χ_n^2 are comfortably below this 20% rejection level.

Program 2.1

Simulates Die Tossing

```
      PROGRAM DICE (OUTPUT,TAPE6=CUTPUT)
C        DICE SIMULATES DIE TOSSING AND COMPARES RELATIVE
C        FREQUENCIES WITH THEIR THEORETICAL VALUES

C        DIFF- ABS(ACTUAL REL FREQ - EXPECTED REL FREQ)
C        ROLL- SET OF 100 SIMULATED TOSSES
C        KCUNT- FREQUENCY COUNTERS FOR EACH OUTCOME
C        CHI- CHI-SQUARED STATISTIC

      REAL DIFF(6)
      INTEGER ROLL(100), KOUNT(6)
      DATA KOUNT /6*0/

C        SIMULATE 100 TOSSES, CCUNT FREQUENCIES AND PRINT THEM
      DO 100 I=1,100
        ROLL(I) = INT( 6.0*RANF(DUMMY) + 1.0 )
        KOUNT( ROLL(I) ) = KOUNT( ROLL(I) ) + 1
100     CONTINUE
      WRITE(6,8C) ROLL
80    FORMAT(1H1,//,22X,"COMPUTER SIMULATION OF DIE TOSSING"///
     1        17X,"OUTCOMES OF THE FIRST 100 TOSSES"//(3X,20I3))

C        SIMULATE 9900 TOSSES, COMPUTE STATISTICS EVERY 500TH
      WRITE (6,81) (I,I=1,6)
81    FORMAT(//"   DIFF(I) IS THE DIFFERENCE BETWEEN THE RELATIVE ",
     1        "FREQUENCY OF"/2X," OUTCOME I",
     2    " AND THE EXPECTED FREQUENCY AFTER N TOSSES"//
     3    "   NUMBER OF",50X,        "CHI-SQUARE"/
     4    "   TOSSES",3X,6( "DIFF(" ,I1, ") "),1X,"STATISTIC"//)

      DO 200 N=101,10000
      ITOSS = INT( 6.0*RANF(DUMMY) + 1.0 )
      KOUNT(ITOSS) = KOUNT(ITOSS) + 1
      IF( MOD(N,500) .NE. 0 ) GOTO 20C
        CHI = C.0
        DO 220 J=1,6
          CHI = CHI + ( FLOAT(KOUNT(J)) - FLOAT(N)/6.0 )**2
          DIFF(J) = ABS( FLCAT(KOUNT(J))/FLOAT(N) - 1.0/6.0 )
220       CONTINUE
```

*The convergence being in the sense that the cumulative distribution functions for the χ_n^2 converge pointwise to the distribution function for χ^2.

```
                        CHI = CHI*6.0/FLOAT(N)
                        WRITE(6,82) N, DIFF, CHI
        82              FORMAT(1X,I6,2X,6F8.5,5X,F8.5)
        200       CONTINUE

                  STOP
                  END
```

COMPUTER SIMULATION OF DIE TOSSING

OUTCOMES OF THE FIRST 100 TOSSES

```
4  6  5  2  3  1  2  2  5  3  1  5  4  1  2  4  1  6  6  5
6  2  1  6  5  4  1  2  5  6  2  4  6  6  3  3  6  6  4  2
6  3  4  3  3  2  5  4  3  3  6  2  5  1  6  5  6  3  2  2
3  1  2  4  4  4  2  3  1  6  5  5  4  1  1  6  3  3  4  6
1  4  6  4  1  5  4  2  3  2  4  5  1  4  4  5  5  4  5  4
```

DIFF(I) IS THE DIFFERENCE BETWEEN THE RELATIVE FREQUENCY OF
OUTCOME I AND THE EXPECTED FREQUENCY AFTER N TOSSES

NUMBER OF TOSSES	DIFF(1)	DIFF(2)	DIFF(3)	DIFF(4)	DIFF(5)	DIFF(6)	CHI-SQUARE STATISTIC
500	.00867	.01733	.00467	.00933	.01667	.00333	2.32000
1000	.01567	.00933	.00367	.00733	.00033	.00233	2.43200
1500	.01267	.00667	.00467	.00467	.00867	.01200	4.20800
2000	.01017	.00433	.00133	.00117	.01183	.00617	3.64000
2500	.01307	.00533	.00333	.00067	.00773	.00267	4.16480
3000	.01233	.00233	.00200	.00333	.00867	.00400	4.74800
3500	.01095	.00162	.00133	.00190	.00990	.00381	5.05257
4000	.01017	.00133	.00158	.00258	.00858	.00392	4.88000
4500	.01089	.00333	.00089	.00378	.01044	.00089	6.87467
5000	.01067	.00327	.00293	.00413	.00933	.00247	7.30000
5500	.00703	.00303	.00406	.00030	.00697	.00067	4.09891
6000	.00417	.00350	.00550	.00217	.00650	.00217	4.01400
6500	.00236	.00297	.00379	.00010	.00518	.00374	2.71692
7000	.00210	.00381	.00262	.00081	.00576	.00167	2.62057
7500	.00000	.00427	.00200	.00013	.00427	.00213	2.02400
8000	.00117	.00167	.00096	.00029	.00333	.00117	.84550
8500	.00039	.00102	.00016	.00016	.00204	.00173	.42729
9000	.00056	.00256	.00067	.00100	.00244	.00211	1.01067
9500	.00088	.00288	.00035	.00312	.00196	.00098	1.35368
10000	.00043	.00287	.00013	.00333	.00063	.00167	1.36280

2.2 SIMULATION OF EXPONENTIALLY DISTRIBUTED OBSERVATIONS

For the __exponential variable__ with parameter λ ($\lambda > 0$), the distribution function is

$$F(x) = \int_0^x \exp(-\lambda y)\, dy = 1 - \exp(-\lambda x), \quad x \geq 0.$$

In the exponential case, $F(x)$ is continuous and therefore takes on every value between 0 and 1. Further, for
each observation of U there is exactly one value X such that
$F(X) = U$:

$$U = F(X) = 1 - \exp(-\lambda X)$$

$$\exp(-\lambda X) = 1 - U.$$

Thus

$$X = -1/\lambda \ln(1 - U).$$

Some slight increase in computational efficiency can be
gained by noting that if U is the uniform variable so is
1 - U and thus

$$X = -1/\lambda \ln(U)$$

must also be exponential.

Program 2.2 begins by printing out 45 observations of
the exponential variable with parameter $\lambda = 2$. These observations are obtained from Algorithm 2.1 as just described.
Next, for N = 100, 500, and 900, N observations of the exponential law are made and the empirical distribution function (defined in Section 1.2) is computed and tabulated.

Plots are presented in Figure 2.3 which compare the
cumulative distribution function of the exponential variable
with the empirical distribution function. A 95% confidence
band is also shown in the plots. The Kolmogorov-Smirnov
statistical theory (Lindgren, Section 6.4.2) assures us
that if the samples really are exponentially distributed,

the empirical distribution function will, with 95% probability, be contained in that band. Conversely, one intuitively expects that if "something is significantly wrong," the empirical distribution function will pass outside the 95% confidence band.

Program 2.2

Simulates Exponential Variate

```
        PROGRAM EXPDIS(OUTPUT,TAPE6=OUTPUT)
C           EXPDIS CALCULATES EXPONENTIAL RANDOM VARIABLES AND
C           EMPIRICAL DISTRIBUTION FUNCTIONS FOR THESE VARIABLES
C           LAMBDA- PARAMETER FOR THE EXPONENTIAL DISTRIBUTION
C           U- A SET OF 50 EXPONENTIAL VARIABLES
C           EDF1-VALUES OF EMPIRICAL DISTRIBUTION FUNCTIONS WITH
C           EDF5,EDF9- 100,500,AND 900 RANDOM VARIABLES
C           TRUE- TRUE VALUES OF EXPONENTIAL DISTRIBUTION
C           X- ORDINATES OF THE DISTRIBUTION FUNCTIONS
C           EDF- SUBROUTINE TO GENERATE EMPIRICAL DISTRIBUTION FUNCTIONS

        REAL LAMBDA, EDF1(50),EDF5(50),EDF9(50),TRUE(50),U(50),X(50)
        DATA LAMBDA /2.0/

        DO 100 I=1,50
          U(I) = -1.0/LAMBDA*ALOG( RANF(DUMMY) )
100       CONTINUE

        WRITE(6,80)  LAMBDA, U
80      FORMAT(1H1,"  EXPONENTIAL VARIABLES     LAMBDA=",G10.4//
     1         10(1X,F7.5))

        CALL EDF(EDF1,100,LAMBDA)
        CALL EDF(EDF5,500,LAMBDA)
        CALL EDF(EDF9,900,LAMBDA)

        DO 200 I=1,50
          X(I) = FLOAT(I)/25.0
          TRUE(I) = 1.0 - EXP( -LAMBDA*X(I) )
200       CONTINUE

        WRITE (6,81) (X(I), EDF1(I), EDF5(I), EDF9(I), TRUE(I),I=2,50,2)
81      FORMAT(///"                    DISTRIBUTION FUNCTIONS"//
     1         3X,"          EMPIRICAL  EMPIRICAL  EMPIRICAL    TRUE"/
     2         3X," X        N=100      N=500      N=900"//
     3         (1X,F5.2,4F11.5))
        STOP
        END
```

```
      SUBROUTINE EDF( VALUE,N,LAMBDA)
C         EDF COMPUTES THE VALUES OF THE EMPIRICAL DISTRIBUTION
C         FUNCTION OF N EXPONENTIAL VARIATES (PARAMATER LAMBDA)
C         AT 1/25, 2/25, ... 50/25.
      REAL LAMBDA, VALUE(50)

      DO 100 J=1,50
        VALUE(J) = 0.0
100     CONTINUE

      DO 200 I=1,N
        U = -1.0/LAMBDA*ALOG( RANF(DUMMY) )
        ISTART = MAX0( 1, INT(25.0*U) )
        IF( ISTART .GT. 50 ) GOTO 200
        DO 220 J=ISTART,50
          VALUE(J) = VALUE(J) + 1.0/FLOAT(N)
220       CONTINUE
200     CONTINUE
      RETURN
      END
```

EXPONENTIAL VARIABLES   LAMBDA= 2.000

```
.27227   .02538   .12016   .60597   .39516 2.53663   .64415   .59266   .18618
1.00907  .09205   .26980 1.15821   .64268   .23866 1.24144   .00483   .01043
.03388   .77537 1.01761   .07390   .09989   .30743 1.98696   .57900   .13382
.62013   .26807   .05100   .06014   .34992   .46256   .06859   .03569   .25919
.05932   .38378   .23258   .40291   .39061   .72430   .13486   .26281   .40615
```

DISTRIBUTION FUNCTIONS

X	EMPIRICAL N=100	EMPIRICAL N=500	EMPIRICAL N=900	TRUE
.08	.28000	.19600	.20111	.14786
.16	.41000	.30200	.32667	.27385
.24	.52000	.39800	.42667	.38122
.32	.62000	.50000	.51667	.47271
.40	.66000	.55800	.58778	.55067
.48	.72000	.61600	.66444	.61711
.56	.77000	.68600	.71111	.67372
.64	.81000	.74400	.75778	.72196
.72	.84000	.77600	.79000	.76307
.80	.86000	.81600	.83444	.79810
.88	.88000	.83600	.85556	.82796
.96	.90000	.86400	.88333	.85339
1.04	.94000	.88000	.90222	.87507
1.12	.94000	.89000	.91000	.89354
1.20	.95000	.89400	.92111	.90928
1.28	.98000	.91000	.93444	.92270
1.36	.98000	.91800	.94111	.93413
1.44	.98000	.93000	.94889	.94387
1.52	.98000	.94000	.95667	.95217
1.60	.99000	.95200	.96333	.95924
1.68	.99000	.96600	.97222	.96526
1.76	.99000	.97000	.97556	.97040
1.84	.99000	.97800	.97889	.97478
1.92	.99000	.98000	.98333	.97851
2.00	.99000	.98800	.98444	.98168

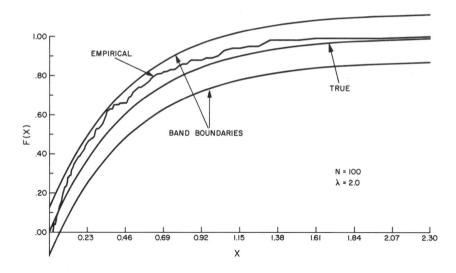

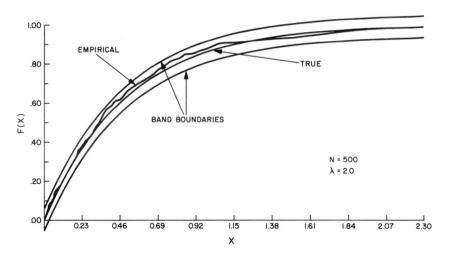

Figure 2.3   Plots of Empiric and Theoretical Distributions

2.3 EFFICIENT SIMULATION OF SOME COMMON PARAMETRIC VARIABLES

Algorithm 2.1 provides a "brute-force" procedure whereby, given any distribution function F and a "good" random number generator, one can obtain observations of the random variable having the distribution F. In practice, however, with the exception of a very few "easy" random variables, Algorithm 2.1 is seldom used directly because, through clever use of numerical and probabilistic facts, one can obtain observations of the common distributions by means of algorithms requiring much less programming and execution time. The clever techniques fall into two broad categories: i) Use of functions of "easy" random variables, and ii) Use of mixtures (to be describes later) in which "easy" random variables are the most probable. By "easy" random variables, we mean of course variables like the exponential that can be obtained with short programs and require but little computer time.

Under i), given a distribution F, one tries to find a computationally fast transformation T and an "easy" random variable (or variables) Y such that if X = T(Y), then X has the desired distribution F. Generation of the binomial variable illustrates this transformation technique.

Generation of binomial observations. The binomial random variable N with parameter (p, n) (p being a number in the unit interval and n a positive integer) has the discrete probability density function

$$P[N = k] = \binom{n}{k} p^k (1 - p)^{n-k}, \ 0 \le k \le n. \qquad (2.4)$$

One may verify (Lindgren, p. 149) that if $N = \sum_{i=1}^{n} X_i$, where
the X_i's are independent Bernoulli variables with parameter
p, then N has the desired binomial distribution. Thus the
sum $\sum_{i=1}^{n} X_i$ serves as the useful transformation of easy
variables $X_1, \ldots, X_n$. The Bernoulli variables (which are
defined to be 1 with probability p and 0 otherwise) are
easily simulated according to Algorithm 2.1 by the rule:
X = 1 if $U \geq (1 - p)$ and X = 0, otherwise. In the remainder
of this section U will consistently denote the uniform
variable (a random number).

For large n, it is computationally more efficient to
make use of the central limit theorem approximation for
sums of Bernoulli variables (Lindgren, p. 143). Thus if
N = Integer Part of $([np(1 - p)]^{1/2}Y + np)$ for Y a standard
normal variable (to be defined later), then for large n,
the distribution of N is "nearly binomial" with parameter
(p, n). The Poisson variable affords an accurate approximation of the binomial variable N when p is small and n is
large (Lindgren, p. 168). Efficient generation of the normal and Poisson variables will be discussed in this section.

Generation of normal observations. The standard normal
variable is determined by the probability density function

$$f(x) = (2\pi)^{-1/2}\exp(-x^2/2), \quad -\infty < x < \infty. \qquad (2.5)$$

A normal variable Y with mean m and standard deviation σ
can be obtained from a standard normal observation X by the
translation $Y = \sigma X + m$. The normal variable is of paramount
importance in probability and statistics. In view of this
importance, we will give three major techniques for standard

normal generation. Unfortunately, the distribution function
of a normal variate cannot be expressed in terms of the elementary functions. For that reason, Algorithm 2.1 can only
be implemented through computationally costly and inexact
numerical integration (quadrature). Thus the normal generators used in practice employ other principles. A particularly simple rule is based on the central limit theorem
(Lindgren, p. 143) which asserts that sums of independent
mean-zero identically-distributed variables, scaled by the
inverse of their standard deviations and divided by the
square root of the number of summands, converge, as the number of summands increases, to the standard normal variable.
Applying this fact to sums of uniform variates, we have the
central-limit algorithm for obtaining an approximately normal variable X_c:

$$X_c = [\sum_{i=1}^{n} U_i - n/2]/(n/12)^{1/2}. \tag{2.6}$$

By taking n to be 12, as is often done, the square root
operation and division can be avoided. This algorithm is
fast, but suffers some drawbacks stemming from the crudeness
of the approximation for manageable values of n. One draw-
back is that the tails of the normal variate can never be
observed: $|X|$ cannot exceed $(n/2)/(n/12)^{1/2}$.

   The Box-Muller algorithm is exact in the sense that it
gives variates which are distributed according to the density (2.5). The Box-Muller (1958) algorithm generates independent standard normal variables X_B and Y_B from random
numbers U_1 and U_2 according to the following rule:

Define $V_j = 2U_j - 1$, $j = 1,2,$

$$S = V_1^2 + V_2^2. \tag{2.7a}$$

Then if $S \geq 1$, call new random numbers U_1 and U_2 and start again. Otherwise, set

$$X_B = V_1(-2(\ln S)/S)^{1/2}$$

$$Y_B = V_2(-2(\ln S)/S)^{1/2}. \tag{2.7b}$$

See Knuth (1969, p. 104) or Fishman (1973, p. 212) for (different) proofs that X_B and Y_B are independent standard normal variates. The accuracy of the Box-Muller method is limited only by the computer word length and the accuracy of the logarithmic and square-root routines. The probability that $S \leq 1$, in (2.7a), is $\pi/4$.

Despite its perfect accuracy, the Box-Muller method has not completely settled the question of efficient normal generation because the functional (log and root) evaluations are relatively time-consuming if they are accurate. In typical numerical analysis and operations research applications, many thousands of normal observations are needed and it is sometimes useful to trade exactness of distribution for speed. The Teichroew method lies in the middle ground between the simplistic central limit algorithm (2.6) and the perfect (but time-consuming) Box-Muller generator. The Teichroew method requires twelve uniform observations (random numbers) U_i, $1 \leq i \leq 12$. Define $R = U_1 + U_2 + \cdots + U_{12} - 6)/4$. Then

$$X_T = ((((a_9 R^2 + a_7)R^2 + a_5)R^2 + a_3)R^2 + a_1)R \qquad (2.8)$$

where

$$a_1 = 3.94984\ 6138,\ a_3 = 0.25240\ 8784,\ a_5 = 0.07654\ 2912$$

$$a_7 = 0.00835\ 5968,\ a_9 = 0.02989\ 9776$$

is the approximation to the standard normal variate. The
motivation of the rule is as follows. Let F_X denote the
distribution function of the standard normal variable, and
F_R the distribution of R above. Then it is evident that the
distribution of $F_R(R)$ is uniform (see problem 8), and so the
distribution of the composition $F_X^{-1}(F_R(R))$ must be the standard normal variable. The Teichroew formula (2.7) gives a
Chebyshev polynomial approximation to $F_X^{-1}(F_R(\bullet))$. As ex-
plained in Muller (1959), the more obvious technique of
making polynomial approximations of F^{-1} itself has not yet
led to satisfactory techniques. Further discussion on the
derivation of the Teichroew method as well as comparison
of the running times, memory requirements, and accuracy of
the three standard normal generators given here are to be
found in Muller's (1959) paper.

Knuth (1969, pp. 105-112) gives details of a method of
Marsaglia and MacLaren for generating standard normal observations. For want of space, we only sketch the ideas behind
this successful but lengthy algorithm. The Marsaglia-
MacLaren generator is based upon a mixture, as per avenue
ii) discussed at the beginning of this section. A mixture
is a sum of distribution functions F_i scaled by constants
p_i such that $p_i > 0$ and $\Sigma\ p_i = 1$. The function F determined

by the mixture is given by

$$F = \sum_i p_i F_i \tag{2.9}$$

and is necessarily a distribution function. In the Marsaglia-
MacLaren algorithm, the distribution function F of the standard normal variate is represented as a mixture (2.9) in
which the distribution functions associated with the larger
p_i's are of very simple form (various uniform distribution
functions). Some of the F_i's associated with small p_i's
are complicated, but they do not need to be used very often.
The mixture (2.9) is used to generate random F-distributed
observations as follows:

   a) Pick an index I, $1 \leq I \leq n$, randomly, so that the
       probability of index k being chosen is p_k.
   b) Generate an observation X(I) according th the dis-
       tribution F_I.

By considering the marginal distribution of X(I), the reader
may verify that X(I) has the desired distribution F (see
Problem 2). If, as in the case of the Marsaglia-MacLaren
generator, the X variates associated with the F_i's having
the larger probabilities p_i of being selected are "easy" to
generate, with high probability, an iteration of the algorithm will require little computation. But the price paid
for having a perfect generator, within the accuracy of computer routines, is that occasionally it will be necessary
to simulate a variate which is not "easy". Knuth (1969)
suggests that the Marsaglia-MacLaren generator is more
efficient than the other perfect generator, the Box-Muller

algorithm. The former is a more complicated algorith,
however, and requires a greater programming effort. We
refer the reader to Knuth (1969) or to Marsaglia, MacLaren,
and Bray (1964) for the extensive details needed to encode
the Marsaglia-MacLaren generator. The implication of Table
2.2 to follow is that the "perfect" Box-Muller algorithm
is sufficiently fast on modern machines that for many purposes, the programming effort required for the Marsaglia-
MacLaren generator is not justifiable.

Let us summarize the results of a computer study intended to validate the schemes just discussed. We generated
sequences of observations on X_c, the central-limit normal
approximation (2.6), the pairs (X_B, Y_B) obtained from the
Box-Muller algorithm (2.7), and observations of X_T, the
Teicherow approximation (2.8). In each case, the Kolmogorov-
Smirnov (K-S) statistic (discussed in Section 1.2) was computed. In Table 2.1, which summarizes the output, one sees
that the performances of the generators were about the same
and each fell within the acceptance region of the K-S
thresholds at the 20% level.

In an additional computer study, we computed 10^4 observations of each of the above normal generators and recorded
the execution times. They are given in Table 2.2 below.
It is curious that on the CDC 6400 computer, the "perfect"
Box-Muller method is significantly faster than the other
algorithms. We attribute this to the fact that two normal
observations are obtained at each iteration, and, apparently,
the CDC functional routines are fast. Muller (1959) found
that the above routines all ran at about the same speed on
the (now nearly extinct) IBM 704.

Table 2.1

K-S Tests for Normality

Method: Number of Samples	Central Limit	Teichroew	Box-Muller
1000[a)	0.022	0.024	0.030
5000[b)	0.0094	0.012	0.015
10000[c)	0.0082	0.0096	0.0055

a)   20% K-S rejection level: 0.034
b)   20% K-S rejection level: 0.015
c)   20% K-S rejection level: 0.011

Table 2.2

Execution Times for 10^4 Normal Observations

Method:	Central Limit	Teichroew	Box-Muller
Time: (in C.P. seconds)	3.147	3.861	1.519

Generation of exponential variables. The exponential variable was defined in Section 2.2, where it was noted that if
U is uniform, X = -ln(U) is exponential with parameter 1.
Marsaglia (1961) as well as Knuth (1969, p. 114) give a
scheme that eliminates the use of the log operation, but
requires use of the two sequences of k values: $\{1 - e^{-j}\}_{j=1}^{k}$
and $\{(e - 1)^{-1}(\Sigma\ 1/v!:\ 1 \leq v \leq j)\}_{j=1}^{k}$. The tables include
only terms having magnitudes commensurate with computer

word length.   Let M be a geometric variable with parameter e^{-1}; that is, $P[M = j] = e^{-j}(1 - e^{-1})$.   The first sequence then gives the distribution of M.   Let N be chosen according to the distribution function given by the second sequence (which is a Poisson variable with the value 0 excluded).   Then the variable

$$X = M + \min (U_1, U_2, \ldots, U_N), \tag{2.10}$$

is exponential with parameter 1.   In (2.10), the U_i's are random numbers.

If $Y = a^{-1}X$, where X is exponential with parameter 1 and a is positive, then Y is exponential with parameter a (See Problem 3).

Generation of gamma variates.   The gamma variable with parameter (α, β), α, $\beta > 0$, has the density function

$$f(x) = \beta^{\alpha}x^{\alpha-1}e^{-\beta x}/\Gamma(\alpha), \quad x \geq 0.$$

$\Gamma(\cdot)$ denotes the gamma function.   One may verify (Problem 3) that if X is gamma $(\alpha, 1)$ and $Y = \beta^{-1}X$, then Y is gamma with parameter (α, β).   For this reason, we devote attention exclusively to $(\alpha, 1)$ gamma distributions.

Some well-known facts about gamma variates will be useful to our discussion:

   A.   If X_j are independent gamma $(\alpha_j, 1)$ variates, $j = 1,2$, then $X_1 + X_2$ is gamma $(\alpha_1 + \alpha_2, 1)$.
   B.   The exponential variate with parameter β is gamma with parameter $(1, \beta)$.

    C. If W is beta* with parameter $(p, 1 - p)$, X is ex-
ponential with parameter 1, and X and W are independent, then

$$Y = WX \qquad\qquad\qquad (2.11)$$

is gamma $(p, 1)$.

Facts A and B are proven in (Wilks, 1962, pp. 170-174)
and Fact C is demonstrated in Fishman (1973, Section 8.2.7).
From consideration of facts A and B above, it is immediate
that if α is integral and $\{X_i\}$ is a sequence of independent
exponential, parameter 1 variates, then

$$Y = X_1 + X_2 + \cdots + X_\alpha$$

is gamma $(\alpha, 1)$. On the other hand, if α is not integral,
then (again from A and B) the gamma variate with parameter
$(\alpha, 1)$ can be represented as

$$Y = X_1 + X_2 + \cdots + X_{[\alpha]} + Z_p,$$

where the X_i's are as above, $[\alpha]$ denotes the integral part
of α, and Z_p is gamma $(p, 1)$ with $p = \alpha - [\alpha] < 1$. From
C, if W is beta with parameter $(p, 1 - p)$ and X is gamma
$(1, 1)$ (that is, exponential with parameter 1), then $Y = WX$
is gamma with parameter $(p, 1)$. This is, in fact, the standard algorithm for generating gamma variates with non-
integral "shape parameter." For this reason, generation of

---

*To be defined in the next subsection.

beta variates, which we now discuss, is particularly
important.

Generation of beta variates. The beta variable with parameter (p, q), p, q > 0, is determined by the probability
density

$$f(x) = (\Gamma(p + q)/\Gamma(p)\Gamma(q))x^{p-1}(1 - x)^{q-1}, \ 0 \le x \le 1.$$

The following procedure is essentially the only exact technique for beta generation to be found in the literature.

1. Choose random numbers U_1 and U_2.
2. Set $S_1 = U_1^{1/p}$, $S_2 = U_2^{1/q}$.
3 Check to see if $S_1 + S_2 \le 1$. If not, return to
   Step 1.
4. The random quantity $S_1/(S_1 + S_2)$ has the beta distribution with parameter (p, q).

In Appendix 2.B, we have provided a proof that the
above algorithm does produce beta variates. The reasons
for including the proof are i) to give the reader exposure
to the type of analysis needed to derive and verify efficient random-variable generators and ii) demonstrations of
the validity of the beta generator are not widely available
in the simulation literature*. It may be seen (Whittaker,
1974) that if q = 1 - p and p is less than 1, then the

---

*The author is aware of only one English-language publica-
tion providing a demonstration, namely (Whittaker, 1974).

probability of "$S_1 + S_2 \leq 1$" is at least $\pi/4$. But the probability of this event can become arbitrarily close to 0 if p or q > 1. Under this latter condition it may be wise to generate independent gamma (p, α) and (q, α) variates X_1 and X_2 and use the fact (Wilks, 1962, pp. 170-174) that under these circumstances, $X_1/X_1 + X_2$ is beta (p, q).

Generation of Poisson variates. Let λ be a positive number. The Poisson variable with parameter λ takes on non-negative integer values and is determined by the discrete probability density function

$$f(n) = P[X = n] = \exp(-\lambda)(\lambda)^n/n!, \ n = 0,1,2,\ldots$$

Elementary probability texts (e.g. Lindgren, Section 3.2.2) show that the Poisson process can be obtained from sequences of exponential distributions. If X_i, i = 1,2,..., are in-dependent exponential variates with parameter 1, and if we define N to be the greatest integer such that $\sum_{i=1}^{N} X_i < \lambda$, then N may be shown to be Poisson with parameter λ. The above rule provides a simulation method for obtaining obser-vations of the Poisson variate with parameter λ. The following method, which appears in Knuth (1969), may be seen to be equivalent to the above procedure for obtaining obser-vations of a Poisson variate with parameter λ. An advantage is that the logarithm needs to be taken only once. As usual, $\{U_i\}$ denotes a sequence of random numbers.

Input. $\{U_i\}$, λ.
   1) Set $p = e^{-\lambda}$, and set i = 1, and S = 1.
   2) Set S to S $\cdot$ U_i.

3) If $S \leq p$, go to 5.

4) Set $i = i + 1$ and go to 2.

5) Set N to $i - 1$.

Output. N, a Poisson variate with parameter λ.

Other variates. Distributions derived from normal variables are conveniently obtained by operating on the normal variable according to the defining equation. Thus, χ^2, the chi-square variable with k degrees of freedom, the t variable with k degrees of freedom, and the F variable with k_1 and k_2 degrees of freedom may be generated according to the respective equations below, where the X_i's and Y_i's denote independent standard normal variates.

$$\text{i)} \quad \chi^2 = X_1^2 + X_2^2 + \cdots + X_k^2,$$

$$\text{ii)} \quad t = Y_1 / ((X_1^2 + X_2^2 + \cdots + X_k^2)/k)^{1/2},$$

$$\text{iii)} \quad F = (k_2/k_1)(X_1^2 + X_2^2 + \cdots + X_{k_1}^2)$$

$$\div (Y_1^2 + Y_2^2 + \cdots + Y_{k_2}^2).$$

From inspection of the density functions, one sees that the chi-square variate with k degrees of freedom is gamma with parameter $(k/2, 1/2)$. For large k, it is faster to use gamma generation techniques for simulating chi-square observations. If X is normal, then e^X is the lognormal variable, another useful progeny of the normal variate.

The geometric variable is an important integer-valued
variate. It is determined by the discreet probability
density function

$$f(n) = P[N = n] = p(1 - p)^n, \; n \geq 1.$$

The parameter p is in the open unit interval. The geometric
variable gives the "waiting time" until a one occurs in a
sequence of independent observations of the Bernoulli variable with parameter p. For values of p close to 1 this defining condition serves as a means for generating geometric
observations from Bernoulli trials, which are trivial to
simulate. For p near 0, the following algorithm in Knuth
(1969) is most effective:

$$N = \text{integer part of } (\ln(U)/\ln(1 - p)) + 1. \qquad (2.12)$$

N, so constructed, is exactly geometric with parameter p.

2.4 BIVARIATE RANDOM VARIABLES AND RANDOM VECTORS

Jointly distributed random variables (X, Y) (often
called bivariate random variables) with density function
$f_{X,Y}(x, y)$ play an important role in probability theory.
In this section, procedures for generating bivariate random
variables are described. Lindgren (Section 2.1) supplies
definitions and discussion concerning these variables. In
particular, a conditional random variable is defined to
have the density function

$$f_{X|Y}(x|y) = f_{X,Y}(x, y)/f_Y(y)$$

provided, of course, that the <u>marginal density</u> $f_Y(y)$
$= \int f_{X,Y}(x, y) \, dx$ does not have value 0. Further, it is
evident from the definition of conditional density that

$$f_{X,Y}(x, y) = f_{X|Y}(x|y)f_Y(y). \qquad (2.13)$$

Let $F_Y(x)$ and $F_{X|Y}(x|y)$ denote the cdf's associated with,
respectively, the marginal density $f_Y(y)$ and the conditional
density $f_{X|Y}(x|y)$. We give below a rule for generating
observations of a bivariate random variable with given density $f_{X,Y}(x, y)$.

ALGORITHM 2.2   BIVARIATE RANDOM VARIABLE GENERATOR.

<u>Input</u>. U_1, U_2, F_Y and $F_{X|Y}$, where U_1, U_2 denote random
numbers.
     1)   Using U_1, compute y, an observation having cdf
         F_Y. (Use Algorithm 2.1 for this step if no better
         way is available.)
     2)   Compute x from U_2 and cdf $F_{X|Y}(\cdot|y)$, y being the
         number obtained in Step 1.

<u>Output</u>. (x, y) is the desired observation.

<u>Verification of Algorithm 2.2</u>. Assume (X, Y) has a (dis-
crete or continuous) density function. By construction, Y
has density f_Y and X has density $f_{X|Y}(x|y)$, y being the out-

outcome of Y. Thus (X, Y) has density $f_{X|Y}f_Y$, which (from equation 2.13) equals $f_{X,Y}$ the desired bivariate density.

Random vectors are the natural generalization to higher dimensions of bivariate random variables. Thus (see Lindgren, Section 2.1 for details.) the probability that a random n-dimensional vector $\underline{X}$ is in a set A is determined by the density function

$$P[\underline{X} \in A] = \int \cdots \int_A f_{X_1, X_2, \ldots, X_n}(x_1, \ldots, x_n) \; dx_1 \ldots dx_n.$$

For each j, $1 \leq j \leq n$, we have a conditional density

$$f_{X_j | X_1, \ldots, X_{j-1}}(x_j | x_1, \ldots, x_{j-1})$$

defined (in terms of marginals) to be equal to

$$\left[f_{X_1, \ldots, X_{j-1}}(x_1, \ldots, x_{j-1}) \right]^{-1} f_{X_1, \ldots, X_j}(x_1, \ldots, x_j).$$

From this, one may verify that

$$f_{X_1, \ldots, X_n}(x_1, \ldots, x_n) = f_{X_1}(x_1) f_{X_2 | X_1}(x_2, x_1) \cdots$$

$$f_{X_n | X_1, \ldots, X_{n-1}}(x_n | x_1, \ldots, x_{n-1}).$$

This leads to the following adaptation of Algorithm 2.2 for generating observations of the random vector $\underline{X} = (X_1, \ldots, X_n)$:

1) Choose x_1 to have the distribution associated with the marginal density f_{X_1}.

2) Define x_j inductively, $1 < j \leq n$, by letting x_j be an observation associated with the (one dimensional) density $f_{X_j|X_1,\ldots,X_{j-1}}(x_j|x_1,\ldots,x_{j-1})$ where x_i $1 \leq i < j$ has already been determined in earlier iterations of this step.

2.5 SUMMARY

The purpose of this chapter has been to give examples using Algorithm 2.1 and alternate more efficient techniques to generate observations of some familiar random variables. The methods of this Chapter, in conjunction with the random number generator given in Chapter 1, will be the basis for the remaining experimentation and methodology in this book.

APPENDIX 2.A

JUSTIFICATION OF ALGORITHM 2.1

The assertion that X, as determined by the Algorithm 2.1, is a random variable having the distribution F(x) is substantiated by the following:

Theorem 2.1:   If U is the random variable uniformly distributed on the unit interval, F(x) is any distribution function and X is the random variable defined by

$$X = \text{minimum} \{y: \ F(y) \geq U\}, \hspace{2cm} (2.14)$$

then X has the distribution function F(x).

Proof: We proceed constructively by determining what the distribution function $\hat{F}(x)$ of the random variable defined in (2.14) must be. Toward this end, we will show that for every real number a, the event [U < F(a)] occurs if and only if the event [X ≤ a] occurs.

[U ≤ F(a)] implies [X ≤ a]

If $u \leq F(a)$, then $a \in \{y: \ F(y) \geq u\} \equiv S$. Therefore the minimum of $S = x \leq a$.

[$X \leq a$] implies [$U \leq F(a)$].

Recalling (2.3), we know that $u \leq F(x)$ Since distribution functions are monotonically increasing, $x \leq a$ implies $F(x) \leq F(a)$.   In summary

$$u \leq F(x) \leq F(a). \hspace{3cm} (2.15)$$

We may conclude that [$U \leq F(a)$] and [$X \leq a$] are equivalent events-- that is, one occurs if and only if the other occurs. Therefore, since X is a random variable _induced_ by the random variable U (see Lindgren, p. 68), they must have the same probability.   Writing this statement in equation form, we have

$$P[U \leq F(a)] = P[X \leq a] = \hat{F}(a).$$

But with the uniform distribution, for all numbers r in the unit interval we have $P[U \leq r] = r$.   Thus

$$P[U \leq F(a)] = \underline{F(a) = \hat{F}(a)},$$

which proves that the distribution function of S, as defined in (2.14), is indeed $F(x)$.

APPENDIX 2.B

JUSTIFICATION OF THE BETA GENERATOR

Recall $S_1 = U_1^{1/p}$, $S_2 = U_2^{1/q}$, where U_1, U_2 are independent uniform variates, conditioned on the event E defined by

$$E = \{U_1^{1/P} + U_2^{1/q} \leq 1\}.$$

Thus the conditional bivariate density of (U_1, U_2) is

$$f_{(U_1,U_2)}(u_1, u_2) = 1/P[E], \text{ all } (u_1, u_2) \varepsilon E,$$

P[E] being the unconditional probability of event E; that is, P[E] is the area of event E. For future reference we now calculate this number. Let $w = (1 - u_2^{1/q})^p$.

$$P[E] = \int_0^1 \int_0^w du_1 \, du_2 = \int_0^1 (1 - u_2^{1/q})^p \, du_2.$$

We make the change of variables, $v = (1 - u_2^{1/q})$. Thus $du_2 = -q(1 - v)^{q-1} dv$ and, reversing the direction of integration

$$P[E] = q \int_0^1 v^p (1 - v)^{q-1} dv = q \frac{\Gamma(p + 1)\Gamma(q)}{\Gamma(p + q + 1)} .$$

Let us define $(x_1, x_2) = (s_1/(s_1 + s_2), s_1 + s_2) \equiv T(u_1, u_2)$, where $s_1 = u_1^{1/p}$, $s_2 = u_2^{1/q}$, as above.

By iterative application of the usual rule for calculating the probability density resulting from bivariate transformations (Lindgren, p. 254), we have

$$f_{(X_1,X_2)}(x_1, x_2) = |\partial(u_1, u_2)/\partial(s_1, s_2)| |\partial(s_1, s_2)/$$

$$\partial(x_1, x_2)| \cdot f_{(U_1,U_2)}(T^{-1}(x_1, x_2)),$$

where the first two terms on the right hand side denote Jacobians of the indicated variables. Easy calculation shows

$$\partial(u_1, u_2)/\partial(s_1, s_2) = pqs_1^{p-1}s_2^{q-1}$$

$$= pqx_1^{p-1}(1 - x_1)^{q-1}x_2^{p+q-2}$$

and $\partial(s_1, s_2)/\partial(x_1, x_2) = x_2$.

Recalling that $f_{(U_1,U_2)}(T^{-1}(x_1, x_2)) = 1/P[E]$, all (x_1, x_2), and integrating out x_2, we have

$$f_{X_1}(x_1) = [pq/(p + q)]x_1^{p-1}(1 - x_1)^{q-1}$$

$$\div qp\Gamma(p)\Gamma(q)/\Gamma(p + q + 1)$$

$$= [\Gamma(p + q)/\Gamma(p)\Gamma(q)]x_1^{p-1}(1 - x_1)^{q-1}, \; 0 \le x_1 \le 1,$$

which is the density of a beta variate with parameter (p, q).

EXERCISES

1) Prove that if U is uniformly distributed, then so is
   1 - U.

2) Let $F_1, F_2, \ldots, F_n$ denote distribution functions and let
   $p_1, p_2, \ldots, p_n$ be positive numbers which add to 1.
   a) Prove that if N is chosen so that $P[N = j] = p_j$ and
      if, then, X is generated to have distribution F_N,
      then

$$F_X(x) = P[X \leq X] = \sum_{i=1}^{n} p_i F_i(x).$$

   b) If, on the average, it takes negligible time to
      generate N, above, and it takes t_j seconds to simu-
      late an F_j distributed observation, generated
      according to the above "mixture" procedure, what is
      the expected time per observation of X?

3) Prove that

   a) If X is exponential with parameter 1 and a is a
      positive number, then $a^{-1}X$ is exponential with param-
      eter a.

   b) If X is gamma with parameter (α, β), then aX is
      gamma with parameter $(\alpha, \beta/a)$.

4) Show that if X and Y are independent random variables,
   then

   $$Z = \max \{X,Y\}$$

   has distribution $F_X(z)F_Y(z) = F_Z(z)$.

5) (Continuation). Let the distribution F be a polynomial
   defined on [0, 1] and having positive coefficients.
   Show how to obtain observations with distribution F by
   the mixture method, using random numbers and the mini-
   mization operation.

6) Prove that the variable X determined in equation (2.10)
   is exponentially distributed.

7) Show that the variable obtained in (2.12) has the geometric distribution.

8) Let R be any continuous random variable and F its distribution function. Show that the induced random
   variable

   $$Z = F(R)$$

   is the uniform variate.

9) Let F(x) $0 \leq x \leq 1$, be a distribution function determined on this domain by a polynomial with positive
coefficients. Use the result of Problem 4 to obtain
an efficient mixture-type generator where each F_j is
simulated by a maximization operation on random numbers.

REFERENCES FOR CHAPTER 2

Box, G., and M. Muller, (1958). "A Note on the Generation
    of Random Normal Deviates." Ann. Math. Statist., 29,
    pp. 610-611.

Cramér, H., (1946). Mathematical Methods of Statistics.
    Princeton University Press, Princeton, New Jersey.

Fishman, G., (1973). Concepts and Methods in Discrete
    Digital Simulation. Wiley, New York.

Knuth, D., (1968). Seminumerical Algorithms. Addison
    Wesley, Reading, Massachusetts.

Marsaglia, G., M. MacLaren, and T. Bray, (1964). "A Fast
    Procedure for Generating Normal Variables." Comm.
    Assoc. Comp. Mach., 7, pp. 4-10.

Muller, M., (1959). "A Comparison of Methods for Generating
    Normal Deviates on Digital Computers." J. Assoc. of
    Comp. Machinery, 6, pp. 376-383.

Whittaker, J., (1974). "Generating Gamma and Beta Random
    Variables with Non-Integral Shape Parameters." Journal
    of the Royal Statistical Society, Series C, 23,
    pp. 210-214.

Wilks, S., (1962). Mathematical Statistics. Wiley, New
    York.

Note. As stated in the preface, Feller and Lindgren are
referenced there, once and for all.

CHAPTER 3

GAMING, RANDOM WALKS, AND LINEAR EQUATIONS

It is
The random
Accumulation

Of triumphs

Which is
So nice.

The Beastly Beatitudes of Balthazar B.,
Donleavy

3.0 BACKGROUND

Most historians agree that the correspondence between
the great 17th Century mathematicians Pierre Fermat and
Blaise Pascal (which is published, in part, in A Source Book
in Mathematics, (Smith, 1959)) concerning a certain gambling
problem marks the birth of the literature of probability

theory.* Moreover, gambling problems continued to provide
impetus and a vehicle for communication in the early development of probability theory in much the same way that physics
spurred mathematical analysis. The historical review by
Maistrov (1974) is a highly readable account of the major
problems and ideas which stimulated the fathers of probability theory. Today the study of two-armed bandits, casinos
(both of these being mathematical idealizations, of course),
and statistical games (i.e., decision theory) are among the
more active and exciting fields of statistical research.
The venerable research monograph How to Gamble If You Must
(Dubins and Savage, (1965)) contains the following quote:

> It is almost always gambling that enables
> one to form a fairly clear idea of a manifes-
> tation of chance; it is gambling that gave
> birth to the calculus of probability; it is to
> gambling that this calculus owes its first
> faltering utterances and its most recent devel-
> opments; it is gambling that enables us to
> conceive of this calculus in the most general
> way; it is, therefore, gambling that one must
> strive to understand. But one should under-
> stand it in a philosophic sense, free from all
> vulgar ideas.

Louis Bachelier

The advantage (aside from the possibility of realizing
whopping big money gains--see E. O. Thorpe, Beat the Dealer,
(1962)) to be gained in the study of gaming is that in this
field, probabilistic questions are posed intuitively and
yet precisely, in a setting that is as devoid as possible
from confusing side issues and ambivalence.

---

*The other contestant for this honor is the works of Cardano,
"the gambling scholar," whose interesting life is narrated
in (Ore, 1953).

In this chapter and the next, we will study certain
simple games whose theoretical properties have been thor-
oughly understood for some while. Our intention in this
experiment is twofold: First we wish to show how easily,
naturally, and convincingly games may be described by simple
probabilistic models. Secondly, we will be able to display
some of the unique didactic strength of the digital computer
by showing how trends may be observed in a large number of
repetitions of a game--trends that seem contrary to intui-
tion, but which we are led to expect for theoretical reasons.

In the closing section of this chapter, we put some of
the game simulation methodology developed in the course of
our analysis to a practical use, namely solving linear equations efficiently. This task is an important impetus to a
branch of simulation technology known as the Monte Carlo
method.

3.1   RANDOM WALKS AND GAMBLERS' FORTUNES

Consider two players, whom we will call A and B, who
are engaged in a sequence of coin tosses. One of them flips
a fair coin at each epoch; if the outcome is "head," A wins
a unit from B; and otherwise, A loses a unit to B. Each
play of this game is then about as simple as a gamble can be.
But as we will see in this chapter and the next, questions
arise naturally concerning the sequence of net fortunes that
are not easy to answer, and the answers turn out to be
surprising, on occasion.

If $X(t)$, $t = 0,1,\ldots$, denotes A's net change in fortune
(from his starting capital) at time t, then $X(t)$ is determined by the rule: $X(0) = 0$, and

$$P[X(t + 1) = n + 1 | X(t) = n]$$

$$= P[X(t + 1) = n - 1 | X(t) = n] = \frac{1}{2}.$$

A graph determined by the $\{X(t)\}$ process defined above is an example of a random walk. In Figure 3.1, we have plotted some simulated random walks.

Random walks, despite being simple to define, are of considerable interest not only for their counter-intuitive properties (some of which will be related in the course of this chapter) but also because other important stochastic processes (most notably Brownian motion and the Poisson process) are limits of random walks. This information is useful both for heuristic purposes and for proving theorems. The details of random walk approximations of Brownian motion are discussed in Chapter 5. Brownian motion constitutes diffusion on a microscopic scale.

Random walks are among the probabilistic constructs known as stochastic processes. A stochastic process is a collection of random variables indexed by a set of real numbers and satisfying certain consistency conditions. Any such set of real numbers is referred to as a time set, and if it is discrete, the stochastic process is sometimes called a time series. L. Breiman's book (Breiman, 1969) constitutes a nice heuristic introduction to the theory and use of stochastic processes and his book (Breiman, 1968) provides a sound introduction to the graduate level mathematics of the subject, as do Karlin and Taylor (1975), Loève (1955), Parzen, (1962), and Bhat (1972).

In this discussion, only some of the many interesting results of Feller, Chapters 3 and 14, on random walks are

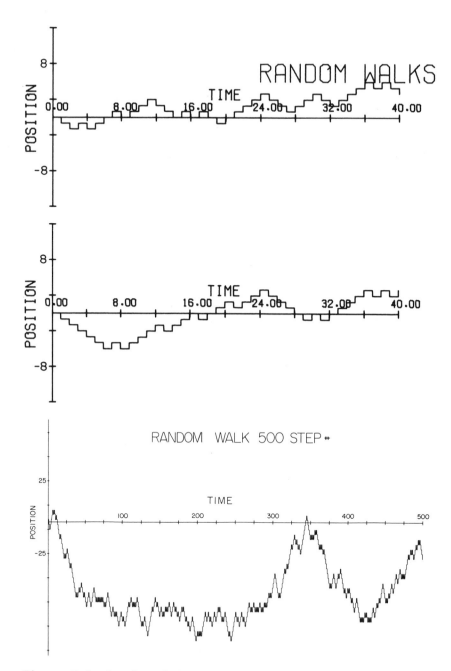

Figure 3.1   Graphs of Fortune Histories of the Coin-toss
              Process

revealed. Feller observes that a number of theoretical results relevant to the coin-tossing problem seem contradictory
to common sense. For example, it is true that if no termination time has been set for the coin-tossing game, no
matter what A's fortune happens to be at any particular
time, the probability is 1 that eventually the players will
be even again. What is surprising is that as soon as the
game starts, it may be a very long time before A and B are
even again. We will derive the fact that the expected time
E[T] of the first zero-crossing equalization is infinite.
Having proved this, it will be interesting, in simulating
the coin-tossing game, to see what happens when we calculate
sample means of a random variable (in this case, T) having
infinite expectation.

In a related but seemingly contradictory development,
as we will see both theoretically and experimentally, walks
in which one player or the other leads about half the time
are much less probable than walks in which one player or
the other is ahead essentially all the time.

Toward laying groundwork for further discussion, we
will demonstrate a few important facts about our "coin-toss"
random walk. In the discussion to follow, we say an
equalization occurs at time n if X(n) = 0.

FACTS ABOUT THE COIN TOSS PROCESS. (a) The probability of
an equalization at time n is given by $\binom{n}{n/2}2^{-n}$ if n is even,
and 0 if n is odd.

(b) The probability is one that infinitely many equalizations will occur in a nonterminating sequence of plays.

(c) The expected time T until the first equalization
is infinite.

Demonstration of the above facts. (a) The probability of
any path $X(t)$, $1 \leq t \leq n$, of possible fortunes is 2^{-n}, the
probability of any particular sequence of n heads and tails.
Thus

$$P[X(n) = 0] = 2^{-n}Q, \tag{3.1}$$

where Q is the number of paths which pass through 0 at time
n. There is a unique path for each arrangement of n/2
heads and n/2 tails. The number of such arrangements is
$\binom{n}{n/2}$, which is the number of ways n/2 heads can be selected from the n coin tosses. Thus $Q = \binom{n}{n/2}$ if n is
even, 0 if n is odd, and (3.1) implies the assertion. More
generally, for $-n \leq k \leq n$, $P[X(n) = k] = \binom{n}{(n+k)/2}2^{-n}$ if
n - k is even and 0 otherwise, as may be seen by extending
the preceding argument.

(b) An important result (Feller, p. 312) on random
walks is that

$$(P[X(n) = 0, \text{ for some } n] = 1)$$

$$\text{if and only if } (\sum_{n \geq 0} P[X(n) = 0] = \infty). \tag{3.2}$$

Applying Stirling's formula (Lindgren, p. 381) to the
factorial terms of $\binom{n}{n/2}$ readily yields that

$$P[X(n) = 0] \simeq (\pi n/2)^{-1/2}, \text{ n even.} \tag{3.3}$$

The divergence of the series $\sum_{n \geq 0} (2n)^{-1/2}$, coupled with
(3.2) and (3.3) imply that certainly at least one equalization occurs. Let E_N, $N = 1, 2, \ldots$, denote the event that at

least N equalizations occur in a non-terminating sequence of
fortunes. We have proved $P[E_1] = 1$. Suppose $P[E_n] = 1$. We
will show this implies $P[E_{n+1}] = 1$, and so by induction,
$P[E_N] = 1$, all N, which implies assertion (b). If $P[E_n = 1]$,
then, with probability 1, in any realization of fortunes,
there is some random time T_n at which $X(t) = 0$ for the <u>nth</u>
time. From the probabilistic description of the $X(t)$ process given at the beginning of this section, the distribu-
tion of $X(T_n + m)|X(T_n)$ is the same as the distribution of
the unconditioned variable $X(m)$. As $X(m)$ is certain to
equal 0 for some $m > 0$, $X(T_n + m)|X(T_n)$ is thereby certain
to be 0 for some m. Thus $P[E_{n+1}|E_n] = 1$. As $P[E_n] = 1$ by
the inductive hypothesis, we have

$$P[E_{n+1}] = P[E_{n+1}|E_n]P[E_n] = 1.$$

(c) For processes such as random walks, the waiting
time T_j between recurrences of any fortune j (T_j is the time
it takes to return to j, given the fortune at the present
time is j) satisfies

$$E[T_j] = \mu(j)^{-1}, \tag{3.4}$$

where

$$\mu(j) = \lim_{n \to \infty} P[X(n) = j].$$

(3.4) is demonstrated in Feller, p. 313.
    The term $\mu(j)$ is the asymptotic probability that the
fortune at any time is amount j. (3.4) implies that the

mean waiting time between recurrences of fortune j is the reciprocal of the asymptotic probability of being in j. This relation is shared by a wide variety of recurrent probabilistic events. For the gambling process, from part (a) of the above listing of facts, we know

$$\mu(0) = \lim_{n \to \infty} \binom{n}{n/2} 2^{-n}$$

$$= \lim_{n \to \infty} (\pi n/2)^{-1/2} = 0.$$

Thus,

Expected time to first equalization

$$= \mu(0)^{-1} = \infty.$$

Let the random variable T be the first equalization time (beyond 0 of course) of the random walk we have been studying. In Program 3.1, we simulate random walks (by simulating a large number of sequences of Bernoulli trials (coin tosses)), each sequence terminating at the time T of the first equalization. The program prints sample means $\bar{T}_n \equiv \sum_{i=1}^{n} T_i/n$ of these equalization times. As the theoretical mean is infinite (fact c above), this sequence of sample means should diverge as n increases. For, it is known (e.g. Breiman, 1968, p. 52) that the condition of finite mean is <u>necessary</u> for convergence of sample means of independent, identically distributed variates.

But even more can be predicted. From Feller, Section III.1, it is seen that if $S_n = \sum_{i=1}^{n} T_i$ denotes the number of tosses required for n such sequences, then for any positive number x, as n becomes large,

$$P[S_n \leq xn^2] \to 2P[Y \geq 1/\sqrt{x}],$$

where Y is the standard normal variable. Thus, roughly speaking, we should anticipate that the sample means $\overline{T}_n \equiv \frac{1}{n} S_n$ ought to grow roughly in proportion to n. In Program 3.1, the coin toss process is simulated and times to succeeding equalizations are averaged. In the printout of Program 3.1, we find that $\overline{T}_{10}$ = (4.6) · 10, $\overline{T}_{100}$ (1.2) · 100, and $\overline{T}_{1000}$ = (1.1) · 1000.

Program 3.1

Simulation of Equalization Times

```
      PROGRAM EQTIME(OUTPUT,TAPE6=OUTPUT)
C         EQTIME FINDS THE AVERAGE TIME TO FIRST EQUALIZATION FOR
C         VARIOUS NUMBERS OF GAMES

C         TBAR-   MEAN TIME TO FIRST EQUALIZATION
C         N-  NUMBER OF GAMES PLAYED
C         ITOSS-  NUMBER OF TOSSES IN CURRENT GAME
C         NTTOSS- TOTAL TOSSES IN ALL GAMES
C         IX- FIRST PLAYERS CURRENT FORTUNE

      WRITE(6,80)
80    FORMAT(1H1,7X" TIMES TO FIRST EQUALIZATION"//
     1        1X,"  NUM OF      AVE TIME TO      ACTUAL TIME TO"/
     2        1X,"  GAMES       EQUALIZATION     EQUALIZATION"/)

      CALL RANSET(23563)
      NTTOSS = 0
      DO 100 N=1,1000
        ITOSS = 0
        IX = 0

C         PLAY TO FIRST EQUALIZATION
120     ITOSS = ITOSS + 1
        IX = IX + 1
        IF( RANF(DUM) .LT. 0.5 ) IX = IX - 2

        IF( IX .NE. 0 ) GOTO 120

        NTTOSS = NTTOSS + ITOSS
```

```
C                 FOR SELECT N CALCULATE AVERAGE TIME TO EQUALIZATION
                  IF( .NOT.( N.LE.10 .OR. N.LE.100 .AND. MOD(N,10).EQ.0 .OR.
         1           MOD(N,100 ).EQ.0 )) GOTO 100
                  TBAR = FLOAT(NTTOSS)/FLOAT(N)
                  WRITE(6,81) N,TBAR,ITOSS
        81        FORMAT(3X,I5,7X,G12.5,1X,I5)
        100       CONTINUE

          STOP
          END
```

TIMES TO FIRST EQUALIZATION

NUM OF GAMES	AVE TIME TO EQUALIZATION	ACTUAL TIME TO EQUALIZATION
1	2.0000	2
2	2.0000	2
3	4.0000	8
4	4.0000	4
5	3.6000	2
6	3.3333	2
7	4.5714	12
8	4.2500	2
9	4.8889	10
10	4.6000	2
20	8.6000	6
30	284.87	10
40	215.95	2
50	173.72	4
60	149.83	2
70	129.77	2
80	142.50	74
90	127.53	6
100	115.96	2
200	1736.9	12
300	1184.9	154
400	901.70	2
500	1097.4	2
600	1212.3	2
700	1047.3	2
800	942.94	12
900	1047.9	2592
1000	1108.4	2

3.2   LONG LEADS IN LONG WALKS

There are some intuitive notions which would seem to suggest that in a long sequence of coin toss plays, each player ought to be in the lead about half the time, and that $X(t)$ sort of oscillates around the t axis. This notion can be expressed more precisely in terms of the law of large numbers, which is now stated formally:

THEOREM 3.1 (THE (STRONG) LAW OF LARGE NUMBERS): Let $\{Z_i\}$ be a sequence of independent random variables with common distribution function F_Z. If the mean of Z_1 is finite, and

$$\overline{Z}_n \equiv 1/n \sum_{i=1}^{n} Z_i,$$

then with probability 1*,

$$\overline{Z}_n \to_n E[Z_1].$$

A statement and proof of the law of large numbers is to be found in Feller, Section X.7.

We apply the law of large numbers to our coin-toss process. Let Z_i be +1 or -1 according to whether the outcome of a coin toss is head or tail. Then

$$X(n) = n\overline{Z}_n = \sum_{i=1}^{n} Z_i$$

where the Z_i's are independent, identically distributed with $E[Z_i] = 0$. From the law of large numbers,

$$X(n)/n = \overline{Z}_n \to_n 0,$$

_____

*This means that the probability of the event "$\overline{Z}_n \to E[Z]$" is 1.

and the average (over time) fortune is certain to converge
to 0 as the number of tosses increases without bound, in
any realization of the coin toss process.

Somewhat surprisingly, in long games, it is much less
likely that each player should be in the lead about half
the time than it is that one player has been ahead during
most of the walk. In fact, the probability is 1/2 that
there has been no equalization during the last half of a
sequence of n plays, regardless of the value n.

Let us briefly study these facts through analysis and
experimentation. A result about our random walk process
which Feller (III.3) calls "the main lemma" is that for n
positive and even,

$$P[X(t) \geq 0, \ 0 < t \leq n] = P[X(n) = 0]. \qquad (3.5)$$

A proof of this consists of showing that for every path
from (0, 0) to (n, 0), there is a uniquely determined path
which lies in the upper half-plane, and conversely, every
path of points $X(t) \geq 0$ has a pre-image among paths passing
through (n, 0). The mapping in question is illustrated in
Figure 3.2. M is defined to be the leftmost minimum of the
path to (n, 0). The portion α' is a reflection of α about
the vertical line passing through M. We have, through this
transformation, established the equivalency of the set of
paths having an equalization at t = n with the set of paths
which are positive from times 1 through n. In view of this
equivalency and the fact that all paths of n steps have the
same probability, (3.5) is verified.

An equivalent statement of the main lemma (see Problem
3) is that for n even,

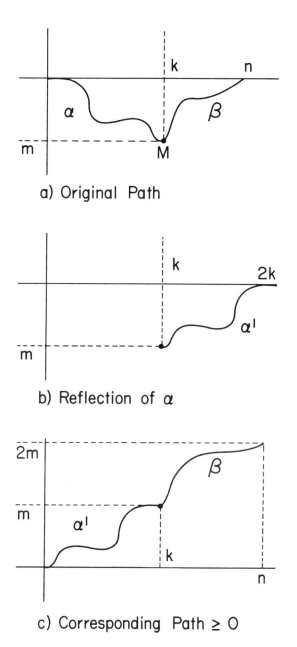

a) Original Path

b) Reflection of α

c) Corresponding Path ≥ 0

Figure 3.2   Construction of Path Transformation

$$P[X(t) > 0, \ 0 < t \le n] = \frac{1}{2}P[X(n) = 0].$$ (3.6)

The following simple application of (3.6) gives (for even k and n) an important conclusion.

$$P[X(k) = 0, \ X(t) \ne 0, \ k < t \le n]$$

$$= P[X(k) = 0]P[X(t) \ne 0, \ k < t \le n \,|\, X(k) = 0]$$

$$= P[X(k) = 0]P[X(t) \ne 0, \ 0 < t \le n - k]$$

$$= u_k u_{n-k}$$ (3.7)

where u_v is defined for even v by

$$u_v \equiv P[X(v) = 0] = \binom{v}{v/2} 2^{-N}.$$

Let $\alpha_{k,n} \equiv P[X(k) = 0, \ X(t) \ne 0, \ k < t \le n]$. Note that by (3.7), $\alpha_{k,n}$ as a function of k is symmetric about n/2. This symmetry clearly implies that with probability 1/2 there has been no equalization during the second half of the walk, regardless of its length. Furthermore, as is demonstrated in Figure 3.3, for the case n = 20, the highest probabilities attach themselves to walks in which the last equalization either occurred very near the end of very near the beginning. Moreover, as Feller (p. 83) proves, $\alpha_{k,n}$ is the probability function of another important event.

Proposition 3.1.   If n and k are even,

$$P[X(t) \ge 0 \text{ for k values of } t \le n]$$

= P[A leads or is even for k time units]

= $\alpha_{k,n}$.

The interpretation of the proposition, in view of the representative graph of Figure 3.3, is that very long or very short leads are the most probable, and that the events in which each player is in the lead about half the time are least probable.

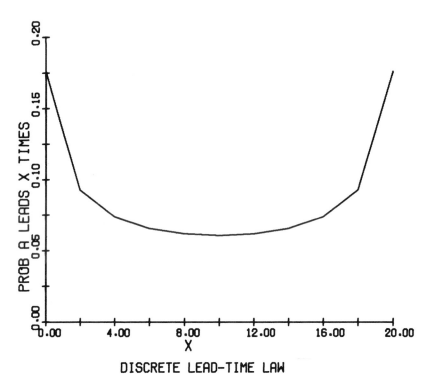

DISCRETE LEAD-TIME LAW

Figure 3.3 Plot of $\alpha_{x,20}$

The function $\alpha_{k,n}$ (which gives the probability of last equalization at k or the probability of being ahead or even k time units out of n) is sometimes called the <u>arc sin law</u> because, as may be quickly derived (Problem 4 or Feller, Chapter III), for large n

$$\sum_{k\leq nx} \alpha_{k,n} \simeq (2/\pi)\ \text{arc}\ \sin(\sqrt{x}). \qquad (3.8)$$

In Program 3.2, we have simulated 1,000 walks of length 50, and computed the relative frequency that one player leads the other for k time units, experimentally confirming that the usual course of events is for one player or the other to be in the lead for the greater portion of the walk. From (3.8), Feller observes that the probability that A or B leads 97.6% of the time is 0.2. In our 1000 simulation runs of a 50 play game created by Program 3.2, the relative frequency that A led between 40% to 60% of the time, was only 0.147, whereas the relative frequency of runs that A leads 0-10% or 91-100% of the time was 0.419.* Yet both of these bands had widths of 20%.

Program 3.2

Simulation Giving Proportion of Plays A Ahead

```
      PROGRAM LEADS (OUTPUT,TAPE6=OUTPUT)
C         LEADS PLAYS 1000 GAMES OF LENGTH 50 EACH AND MAKES A
C         TABLE OF RELATIVE FREQUENCIES OF THE  OF TIMES ONE PLAYER IS AHEAD
C
C         RELFRQ- RELATIVE FREQUENCIE THAT A PLAYER IS AHEAD FOR (I-1) MOVES
C         IX- PLAYER A'S FORTUNE
C         LEADA-  NUMBER OF TIMES PLAYER A IS AHEAD
C         LEADB-  NUMBER OF TIMES PLAYER B IS AHEAD
```

*The theoretical probabilities of these events are 0.130 and 0.400, respectively.

```
                REAL RELFRQ (51)
                DATA RELFRQ /51*0/

                DO 100 N=1,1000
                  LEADA = 0
                  LEADB = 0
                  IX = 0

                DO 120 I=1,50
                  U = RANF (DUMMY)
                  IF( U .GE. 0.5 ) IX = IX + 1
                  IF( U .LT. 0.5 ) IX = IX - 1
                  IF ( IX .GT. 0 ) LEADA = LEADA + 1
                  IF ( IX .LT. 0 ) LEADB = LEADB + 1
    120           CONTINUE

                RELFRQ( LEADA + 1 ) = RELFRQ( LEADA + 1) + 0.0005
                RELFRQ( LEADB + 1 ) = RELFRQ( LEADB + 1) + 0.0005
    100         CONTINUE

                WRITE(6,80) RELFRQ(1) , (( I,RELFRQ(I+1),I=J,50,10),J=1,10)
    80          FORMAT(1H1,4X," RELATIVE FREQUENCIES THAT A PLAYER"/
       1                 5X," IS AHEAD N MOVES IN 50 MOVE GAMES"///
       2                 5X,5( " N    REL FREQ",3X)/
       3                 5X,5( "      AHEAD N ",3X )/
       4                 5X,5( "      MOVES   ",3X)//
       5                 5X," 0",F7.4/ 5(5X,I2,F7.4,3X) )
                STOP
                END
```

RELATIVE FREQUENCIES THAT A PLAYER
IS AHEAD N MOVES IN 50 MOVE GAMES

N	REL FREQ AHEAD N MOVES	N	REL FREQ AHEAD N MOVES	N	REL FREQ AHEAD N MOVES	N	REL FREQ AHEAD N MOVES	N	REL FREQ AHEAD N MOVES
0	.1090								
1	.0575	11	.0190	21	.0130	31	.0150	41	.0140
2	.0270	12	.0165	22	.0115	32	.0145	42	.0140
3	.0290	13	.0150	23	.0140	33	.0120	43	.0160
4	.0205	14	.0110	24	.0070	34	.0120	44	.0135
5	.0195	15	.0140	25	.0130	35	.0175	45	.0185
6	.0195	16	.0155	26	.0145	36	.0130	46	.0225
7	.0180	17	.0135	27	.0150	37	.0120	47	.0210
8	.0175	18	.0145	28	.0100	38	.0160	48	.0335
9	.0185	19	.0115	29	.0160	39	.0155	49	.0235
10	.0130	20	.0180	30	.0150	40	.0140	50	.0555

3.3 MARKOV CHAINS AND LINEAR EQUATIONS

Random walks such as exemplified by our coin toss
histories are a subclass of an important collection of time
series known as Markov chains. In this section, we will
find practical use for Markov chains. A (stationary)
Markov chain is characterized by a set S (called the state
set) and a function $P(x; y)$ of two variables on S such that

$P(x; y) \geq 0$, all x, $y \in S$

and

$$\sum_{x \in S} P(x; y) = 1, \text{ all } y \in S.$$

P is termed a (stationary) <u>transition function</u> and, for fixed state y, is a probability distribution function in x. An initial state $x(0) \in S$ having been specified, $P(x; y)$ determines a random sequence as follows:

$X(1)$ is chosen randomly according to $P(\cdot; x(0))$.

$X(1)$ having been observed, $X(2)$ is chosen randomly according to $P(\cdot, X(1))$.

In general, $X(j)$ having been observed, $X(j + 1)$ is chosen randomly according to the discrete probability density function $P(\cdot; X(j))$, and so forth. Random sequences $\{X(t)\}$ having this structure are termed Markov chains.

In what follows, S will be a set of real numbers. For $s \in S$, observations having the law $P(\cdot; s)$ can be obtained, by authority of Algorithm 2.1, through the rule $X = s'$, where s' is the least member of S such that

$$\sum_{s_1 \leq s'} P(s_1; s) \geq U,$$

U being a random number. Thus, if S is a set containing a small to moderate number of elements, one may simulate a Markov chain $X(t)$, $0 \leq t \leq N$, with a short computer program and modest execution time.

Our intention in this section is to show how Markov chains can be used to solve simultaneous <u>linear equations.</u> Such equations have the form

$$\sum_{j=1}^{n} a_{ij}z_{j} = b_{i}, \ 1 \leq i \leq n \qquad\qquad (3.9)$$

where the a_{ij}'s and b_{i}'s are given constants and the z_{j}'s are the numbers to be found. It is more convenient for our purposes to use matrix notation. We write (3.9) as $Az = b$, and suppose, for now, that A is expressible as

$$A = I - H \qquad\qquad (3.10)$$

where I is the identity matrix and H has the property that all its eigenvalues are less than 1 in magnitude. Then

$$z = A^{-1}b = (I - H)^{-1}b = Ib + Hb + H^{2}b + \ldots + H^{n}b + \ldots,$$

which is known to be convergent if our assumption about the eigenvalues of H holds.

Let us use "path" to designate a sequence $\underline{X}(t)$ of t integers $X(1),\ldots,X(t)$, where the $X(i)$'s take on values in the set $\{1,2,\ldots,n\}$, n being the order of the linear system (3.10). Let P be a probability function which attaches positive probability to every possible path $\underline{X}(t)$ of all possible lengths $t = 0,1,2,\ldots$. That is.

$$P(\underline{X}(t)) > 0, \ \text{all } t \geq 0 \ \text{and all } \underline{X}(t).$$

Note that in this construct, the path length t as well as
the path $\underline{X}(t)$ is random.

A Monte Carlo solution to the linear equation (3.10)
may be obtained by using the following clever device attri-
buted by Forsythe and Liebler (1950) to Ulam and von Neumann:
Paths $\underline{X}(t)$ are simulated according to the distribution $P(\cdot)$
above, and the quantity

$$V(t; x(0)) = (\prod_{i=1}^{t} h_{X(i-1),X(i)}) b(X(t))/P(\underline{X}(t)) \quad (3.11)$$

is computed, the h_{ij}'s being the coordinates of the H matrix
in (3.10). $x(0)$ is a fixed initial coordinate number,
$1 \le x(0) \le n$. This simulation is repeated m times to obtain
$V_1(t_1; x(0)),\ldots,V_m(t_m; x(0))$ and the sample average

$$\overline{V}_m(x(0)) = \sum_{i=1}^{m} V_i(t_i; x(0))/m$$

is computed. It is easy to verify that $E[V(t; x(0))]$
$= (A^{-1}b)_{x(0)} = z_{x(0)}$, where the subscript indicates the
coordinate number of the vector in parentheses. Therefore,
by the law of large numbers, with a probability of 1,

$$\overline{V}_m(x(0)) \to_m (A^{-1}b)_{x(0)} = z_{x(0)}.$$

The fact that $E[V(t; x(0))] = z_{x(0)}$ is readily checked:

$$E[V(t; x(0))] = \Sigma \frac{(\prod_{i=1}^{t} h_{X(i-1),X(i)})b_{X(t)}}{P(X(1),\ldots,X(t))}$$

$$\times P(X(1),\ldots,X(t)),\tag{3.12}$$

where the summation is taken over all possible paths of all possible lengths. The probabilities in (3.12) may be cancelled, giving

$$E[V(t; x(0))] = (b)_{x(0)} + (Hb)_{x(0)} + \cdots + (H^{n}b)_{x(0)} + \cdots$$

$$= ((I - H)^{-1}b)_{x(0)} = z_{x(0)}.$$

Forsythe and Liebler (1950) and others have employed Markov chains for generating random paths. In addition to the index states $\{1,2,\ldots,n\}$, an additional "stopping state" is used.

Let $t + 1$ be the time at which the "stopping state" (which we will denote below by W) occurs. Then for Markov chain paths, (3.11) may be rewritten

$$V(t; x(0)) = b(x(t))[\prod_{i=1}^{t} h_{X(i-1),X(i)}/P(X(i); X(i - 1))]$$

$$/P[W; X(t)]\tag{3.13}$$

In Program 3.3, we solve a linear equation of dimension 170. In this program the probability of passing to the "stopping state" W is chosen, independently of state, to be 0.1. Thus the stopping time t is a geometric variable. Also, we have set $P(x, y) = 0.9/n$, all x and y. As a consequence, our Markov chain is a sequence of independent

variates, uniformly distributed on the index set $\{1,\ldots,n\}$.
The coefficients of H were set equal to different random
numbers, each divided by the dimension n of the system of
equations. Thus $H = \{h_{i,j}\}$, where

$$h_{i,j} = U_{i,j}/n$$

where the U_{ij}'s are random numbers. This assures, by virtue
of the Gerschgorin theorem (Isaacson and Keller, 1966,
p. 135), that the eigenvalue magnitude condition will be
satisfied.

In our computations, we employ the obvious device of
making the same random walk $X(1),\ldots,X(t)$ serve for an observation of the estimate of each solution coordinate z_i in
(3.9): At each iteration, our program simulates a chain
$X(1),\ldots,X(t)$ and for each coordinate number $i = 1,\ldots,n$
we set $x(0) = i$ and record the z_i estimate

$$V(t,\ i) = b(t)[\ \prod_{i=1}^{t} h(_{X(i-1),X(i)}/\frac{(0.9)}{n})]/0.1.$$

For purposes of comparison, the solution to the linear
equation (3.9) has been computed also by Gauss elimination.
In the program printout, one sees that the computational
time of the Monte Carlo algorithm is about a sixth of the
time required by the Gauss routine. The output provides the
sample variance (or sample square error) of each observation
of $V(t,\ i)$, $i = 1,2,\ldots,n$. If the sample variance is divided
by the number of chains simulated (in our case, 1000), then
an estimate of the expected squared error is obtained.

Program 3.3

Markov Chain Solution of a Linear Equation

```
          PROGRAM LINEAR (OUTPUT=80,TAPE6=OUTPUT,TAPE5=80)
C             LINEAR APPROXIMATES THE SOLUTION OF THE LINEAR SYSTEM
C             AZ = (I - H) Z = B  BY THE ULAM-VON NEUMANN METHOD

C             N- DIMENSION OF THE SYSTEM
C             M- NUMBER OF APPROXIMATIONS AVERAGED
C             H- MATRIX WHICH DEFINES THE SYSTEM
C             B- RIGHT HAND SIDE OF THE SYSTEM
C             VM- AVERAGE OF APPROXIMATIONS TO THE SOLUTION
C             Z- EXACT SOLUTION GOT BY GAUSS ELIMINATION
C             VSQ- SAMPLE VARIANCE
C             T- LENGTH OF A MARKOV CHAIN
C             PRSTOP- STOPPING PROBABILITY
C             XIM1,XI- SUCCESSIVE ELEMENTS OF A MARKOV CHAIN
C             V- THE PRODUCT OF H'S USED TO APPROXIMATE Z(J)

C             GAUSS- SUBROUTINE  WHICH SOVES THE SYSTEM BY GAUSS
C                    ELIMINATION
          COMMON H(170,170), B(170), VM(170), Z(170), ERRSQ(170)
          INTEGER T, XI, XIM1, X1
          DATA N,M,PRSTOP / 170,1000,0.1/

C             GENERATE THE SYSTEM
          DO 100 J=1,N
          DO 100 I=1,N
          H(I,J) = RANF(DUMMY)/FLOAT(N)
100       CONTINUE
          DO 120 I=1,N
          B(I) = RANF(DUMMY)
          VM(I) = 0.0
          VSQ(I) = 0.0
120       CONTINUE

          CALL GAUSS(GTIME, N)


          TIME = SECOND(DUMMY)
          DO 200 K=1,M
C             A PRODUCT OF H'S FOR A PARTICULAR CHAIN IS USED TO
C                GET AN ESTIMATE FOR EACH COMPONENT
          T = ALOG( RANF(DUMMY) )/ALOG(1.0 - PRSTOP)
          IF ( T .GT. 0 ) GOTO 220
             DO 210 J=1,N
             VM(J) = VM(J) + B(J)/PRSTOP
             VSQ(J) = VSQ(J) + ( B(J)/PRSTOP )**2
210          CONTINUE
          GOTO 200
220       CONTINUE
          V = 1.0
          X1 = FLOAT(N)*RANF(DUMMY) + 1.0
          XI = X1
          XIM1 = X1
          IF ( T .EQ. 1 ) GOTO 250
             DO 230 I=2,T
             XI = FLOAT(N)*RANF(DUMMY) + 1.0
             V = V*H(XIM1,XI)*FLOAT(N)/(1.0 - PRSTOP)
             XIM1 = XI
230          CONTINUE
250       V = V*B(XI)/PRSTOP
          DO 260 J=1,N
          TEMP = V*H(J,X1)*FLOAT(N)/(1.0 - PRSTOP)
          VM(J) = VM(J) + TEMP
          VSQ(J) = VSQ(J) + TEMP**2
260       CONTINUE
200       CONTINUE
```

```
C           AVERAGES ARE CALCULATED
            DO 300 J=1,N
            VM(J) = VM(J)/FLOAT(M)
            VSQ(J) = ( VSQ(J) - FLOAT(M)*VM(J)**2 )/FLOAT(M - 1)
300         CONTINUE
            TIME = SECOND(DUMMY) - TIME

C               NOTE VALUES WERE FOUND FOR EACH COMPONENT BUT PRINTED
C               ONLY FOR EVERY 5-TH COMPONENT
            WRITE(6,80) (J,VM(J),Z(J),VSQ(J), J=1,N,5)
80          FORMAT(1H1," COMPONENT    APPROX    TRUE      SAMPLE"/
       1            1X,"              SOLUTION  VALUE     VARIANCE"//
       2            ( 5X,I3,5X,3G11.5) )

            WRITE(6,81) TIME, M
81          FORMAT(//10X,F10.2," SECONDS FOR MARKOV CHAIN METHOD"/
       1            12X,I6, " TRIALS AVERAGED")


            WRITE(6,83) GTIME
83          FORMAT(//10X,F10.2," SECONDS FOR GAUSS ELIMINATION")
            STOP
            END
```

COMPONENT	APPROX SOLUTION	TRUE VALUE	SAMPLE VARIANCE
1	.74753	.74452	1.9783
6	1.0400	.97618	3.7124
11	.79624	.78638	2.1886
16	.55610	.58133	1.3591
21	.88284	.86837	2.5493
26	.75602	.72668	2.1200
31	.64228	.64703	1.8351
36	.50574	.55038	1.6610
41	1.3892	1.3011	7.4719
46	1.0767	.97202	4.0294
51	.97684	.90834	3.3815
56	.83562	.84490	2.2650
61	.45327	.48666	1.5688
66	.99311	.93959	3.1787
71	1.2959	1.1822	6.0763
76	.89950	.82068	2.8153
81	1.5329	1.4060	9.0243
86	.86453	.80714	2.7279
91	1.0438	.99160	3.7534
96	1.4900	1.3397	8.3861
101	.76491	.76916	1.9295
106	.70307	.69174	1.8577
111	.40440	.46511	1.1748
116	1.3539	1.2833	7.4580
121	.76820	.74087	1.9804
126	1.4848	1.3723	8.8357
131	1.1766	1.0982	4.7320
136	1.2604	1.1798	5.6540
141	1.2565	1.1756	5.9242
146	.78132	.77134	2.0536
151	.50834	.52501	1.6821
156	1.0565	1.0135	3.7710
161	1.3414	1.2408	6.1597
166	1.3500	1.2528	6.9574

```
           6.39 SECONDS FOR MARKOV CHAIN METHOD
           1000 TRIALS AVERAGED


           38.03 SECONDS FOR GAUSS ELIMINATION
```

Recall that our simulation solution to the linear
equation (3.9) is predicated on the assumption that the
eigenvalues of H are less than 1 in magnitude. But, as
pointed out by Halton (1970), one may always "prepare" the
equation to satisfy this constraint, as follows:

If $Az = b$, then for any matrix C, $CAz = Cb$, and
$H = I - CA$. If A is nonsingular, by proper choice of C,
any H can be obtained. One may proceed by choosing an
easily invertible matrix C to satisfy the eigenvalue constraint, and then solve the system

$$(I - H)z_1 = A_1 z_1 = b_1,$$

where $A_1 = CA$, $b_1 = Cb$. Finally, the solution to the original problem is $z = C^{-1} z_1$.

The error analysis methodology (Halton, 1970, and
Curtiss, 1956) currently available for Monte Carlo solution
of linear equations requires not only that the eigenvalues
of H have magnitudes less than one, but that this same
eigenvalue condition hold for the matrices H^+ and K, where

$$(H^+)_{ij} = |h_{ij}| \quad \text{and} \quad (K)_{ij} = h_{ij}^2/P(j; i),$$

$$1 \leq i, j \leq n.$$

Halton (1970) expressed doubts about whether, for an arbitrary matrix A, there always exists a Markov chain and a
matrix which enables the linear equation to be prepared to
satisfy the eigenvalue condition for H^+ and K. But he
supplies a construction of C for certain classes of matrices.
Supplementary discussion is to be found in (Halton, 1967).

Theorem 2 of Forsythe and Liebler (1950) gives a formula
yielding the expected estimation error for a single path:

$$E[(V(t; x(0)) - z_{x(0)})^2]$$

$$= \sum_{j=1}^{n} (T_{ij}P[stop|j]^{-1} - ((I - H)^{-1})_{ij}^2)b_j^2.$$

where $T \equiv (I - R)^{-1}$, R being the matrix with coordinates
determined by $R_{ij} = A_{ij}^2 P[j;i]$. One can conclude that if
the order of magnitude of the sum in (3.13) remains the
same, the amount of computation by the preceding method and
its variants increases proportionately to n^2, where n is
the order of the linear system, whereas the computation com-
plexity of the Gauss elimination method and its variants
is proportional to n^3 (Isaacson and Keller, 1966, p. 36).
Thus for large enough linear systems, the Monte Carlo
approach must be most efficient. More information and
justification of this assertion is to be found in Curtiss
(1956). Halton (1970) describes linear equations as "one
of the most fruitful fields for the application of the Monte
Carlo method."

We refer the reader to Section 2.3 of Halton (1970) for
further developments and references to the Markov chain
solution of linear equations, and the extension of these
techniques to the solution of eigenvalue problems and
integral equations.

EXERCISES

1) A roulette wheel has thirty-eight divisions, of which 18 are black, 18 are red, and two are green. Assume the gambler bets one unit at each play and that the wheel is balanced correctly. Simulate the sequence of fortunes under the assumption (a) that the gambler bets at even odds on red, (the probability of his winning is 18/38), and also simulate a sequence of fortunes while assuming (b) that the gambler bets on green at odds of 17/1. That is, if the outcome is green, the gambler receives 17 units. Otherwise, he loses a unit. Note (as in Problem 2) that in this process, the probability of an equalization occurring is less than one. The reader may wish to consult Epstein (1967) for discussion from a probabilistic viewpoint of roulette as well as other games of chance.

2) Use the law of large numbers (Theorem 3.1) to show that if the coin is not fair (i.e., $P[X_{n+1} = k + 1 | X_n = k] \neq 1/2$) then with probability of one, at most finitely many equalizations occur in a random walk.

3) Show that equation (3.6) is equivalent to the "Main Lemma." (Hint: Consider the random walk process beginning at (1, 1).)

4) Demonstrate equation (3.8). (Hint: Use Stirling's formula on the combinatorial terms and, after some algebra, approximate the Riemann sum by a standard integral.)

5) (The Petersburg Paradox) Suppose one flips a fair coin until the first tail appears. If this occurs at the nth toss, then the payoff is $\$2^n$. Analytically, find the expected return from this game. Simulate a few repetitions of the game by computer. Are the average payoffs converging? See Feller, p. 252, for more discussion on this famous problem.

6) Show that the coin toss process discussed in this chapter is a Markov chain. What is the transition function for the Markov chain model?

7) Use the Monte Carlo technique to solve a linear equation of dimension compatible with your computer's memory. Make sure the H, in your equation, satisfies the eigenvalue condition. Experiment with different Markov chain transition functions and with different stopping rules. Use a conventional numerical method to check the accuracy of your various estimates.

8) Let $\{X(i)\}_{i=1}^{\infty}$ be a nonterminating Markov chain with values on the indices of a linear equation, as per Section 3.3. Show that

$$V'(X(0)) \equiv b_{X(0)}$$

$$+ \sum_{i=1}^{\infty} \sum_{j=1}^{i} [h_{X(j-1),X(j)}/P(X(j), X(j-1))]b_{X(i)}$$

is an unbiased estimator of $z_{X(0)}$. Wasow (1952) and Halton (1967) have studied truncations of the estimator V'.

REFERENCES FOR CHAPTER 3

Breiman, L., (1969). Probability and Stochastic Processes,
    with a View Toward Application. Houghton Mifflin,
    Boston, Mass.

Breiman, L., (1968). Probability. Addison Wesley, Reading,
    Mass.

Curtiss, J., (1956). " A Theoretical Comparison of the
    Efficiencies of Two· Classical Methods and a Monte Carlo
    Method for Computing One Component of a Solution of a
    Set of Linear Algebraic Equations." Symposium on Monte
    Carlo Methods, H. Meyer, Ed. Wiley, New York,
    pp. 191-233.

Dubins, L., and L. Savage., (1965). How to Gamble if You
    Must. McGraw-Hill, New York, New York.

Epstein, R., (1967). The Theory of Gambling and Statistical
    Logic. Academic Press, New York.

Forsythe, G., and R. Leibler, (1950). "Matrix Inversion by
    a Monte Carlo Method." Mathematical Tabs. Aids Comput.,
    4, pp. 127-129.

Halton, J., (1967). "Sequential Monte Carlo (Revised)."
    Technical Summary Report No. 816, Mathematics Research
    Center, University of Wisconsin, Madison, Wisconsin.

Halton, J., (1970). "A Retrospective and Prospective Survey
    of the Monte Carlo Method." SIAM Review, 12, pp. 1-63.

Isaacson, E., and H. Keller, (1966). Analysis of Numerical
    Methods. Wiley, New York.

Karlin, S., and H. Taylor, (1975). A First Course in Stochastic Processes. Second edition, Academic Press,
New York.

Loeve, M., (1955). Probability Theory. Van Nostrand,
Princeton, New Jersey.

Ore, O., (1953). Cardano, the Gambling Scholar. Princeton
University Press, Princeton, New Jersey.

Parzen, E., (1962). Stochastic Processes. Holden-Day,
San Francisco.

Maistrov, L., (1974). Probability Theory, a Historical
Sketch. Academic Press, New York, New York.

Smith, D., ed. (1959). A Source Book in Mathematics.
Dover, New York, New York.

Thorpe, E. O., (1962). Beat the Dealer. Blaisdell Publishing Company, New York, New York.

Wasow, W., "A Note on the Inversion of Matrices by Random
Walks." Math. Tab. Aids Comput., 6, pp. 78-81.

CHAPTER 4

GAMBLER'S RUIN WITH EXTENSIONS TO INVENTORY CONTROL

"Well wagered," Burlingame said merrily, "and
I'll add more: who loses must not merely
walk, but walk behind Old Roan there, that
ever gets the bumbreezes near midmorning.
'Twill add a spice to the winner's victory!"

Done," agreed the poet. "I had in sooth
observed the mare was flatulent."

Sot-Weed Factor,
Barth

4.0   BACKGROUND

An idealization suitable for our analysis of the situation of a gambler in a casino is that a number of different
subfair games (games in which the expected return from a
play is negative) are available, and the gambler is determined to play until either he or the casino runs out of
money. If the gambler has the less ambitious goal of

attaining some gain Q rather than "breaking the bank," just
assume that the casino's capital is Q and the situation is
mathematically equivalent to the "break the bank" problem
as far as the gambler is concerned. The simplification is
made during most of this chapter that there is but one gamble available, and the probability of the gambler's winning
it is p. The return is plus or minus 1, according to
whether or not he wins the gamble. The classical gambler's
ruin problem is the problem of determining the probability
that the gambler initially having a capital of x units will
fail to break the bank and lose his capital at a casino
initially having a - x units of capital. The gambler's ruin
problem is studied analytically and experimentally in
Section 4.1.

The solution of the gambler's ruin problem is instructive in that one can see that with fair games (p = 1/2), the
fact that the casino has more capital than the gambler gives
it no advantage; the gambler's expected loss is always zero.
This gambler's ruin result is subsumed in a very general
class of theorems of decision theory which we shall call the
principle of conservation of fairness, the topic of discourse in Section 4.2. Roughly, the principle says that in
a fair casino (that is, one offering only fair gambles), no
matter what games are available to the gambler and no matter
what strategies the gambler may follow, his expected loss
is zero. This principle is not generally true if the payoff
during a play is a random variable with infinite variance
(See Feller, p. 249f, for a counterexample).

Thus far, our gambles have been the same at each play.
In Section 4.3, we allow variable bet amounts and explore
(analytically and through simulation) the effectiveness of

different betting strategies. This study is motivated by
the central problem in the treatise How to Gamble If You
Must (Dubins and Savage, 1965), which demonstrates the opti-
mality of "bold" betting strategies for avoiding "gambler's
ruin" in subfair casinos.

It will be seen in Sections 4.4 and 4.5 that the gambler's ruin process is a prototype of more complicated
problems in operations research. Specifically, in actuary
science and inventory control respectively, the random insurance claims and product demand correspond to the outcome
of a play of the game. An insurance company is ruined if
the claims exceed the current fortune. In inventory control,
running out of stock is equivalent to ruin, at least until
the next shipment. In the final portion of the chapter, we
describe some of the bonds connecting gambler's ruin processes with sequential methods of statistics.

4.1 THE PROBABILITY OF A GAMBLER'S RUIN

Chapter 4 is devoted to studying gambler's ruin problems and their generalizations, which are an interesting
aspect of the gambling process introduced in Chapter 3.
Let us examine gambler's ruin in terms of the earlier random
walk problem. Suppose the gambler's fortune initially is
x and the house has an initial capital a - x. The event
that the gambler is ruined is the set of walks which begin
at (0, 0) and reach the ordinate -x before reaching ordinate
a - x.

In Figure 4.1 below, we give plots of two simulated
random walks, showing the absorbing barriers. In these
two plots, the probability p of winning a play was,

respectively, 0.65 and 0.43, and absorbing barriers were
at ±20.

We have observed that a path describing a gambler's
change in fortune is called a "random walk." In random
walk terminology, the horizontal lines through (0, -x) and
(0, a - x) are called <u>absorbing barriers</u>, and being ruined
or breaking the bank are equivalent to being absorbed, i.e.,
touching an absorbing barrier. Viewing the ruin process as
a random walk with absorbing barriers induces a number of
further questions about the gambler's ruin. We shall here
restrict our attention to the question: "What is the
probability of ruin?"

Let the integer x be the fortune of the gambler, and
a - x be the capital of the casino. p is the probability
that the gambler wins a play. q_x is the probability that,
given the gambler has fortune x, he will eventually be
ruined. (Thus $q_0 = 1$, $q_a = 0$.)

If $x \neq 0$ or a, then at least one more play remains, and
the fact that either a win or a loss must result from that
gamble yields the relationship

$$q_x = pq_{x+1} + qq_{x-1}, \quad 0 < x < a, \tag{4.1}$$

in which $q \equiv 1 - p$.

Some readers may find the following justification of
(4.1) helpful. A special case of the total probability
rule (Lindgren, p. 37) is

$$P[E] = P[E|F]P[F] + P[E|F^c]P[F^c] \tag{4.2}$$

F^c being the complement of the event F. Let E be the event

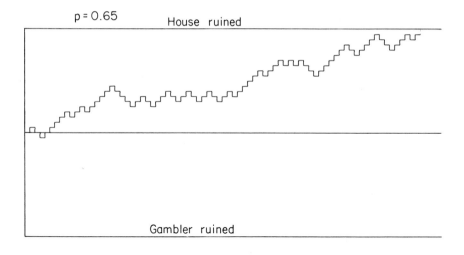

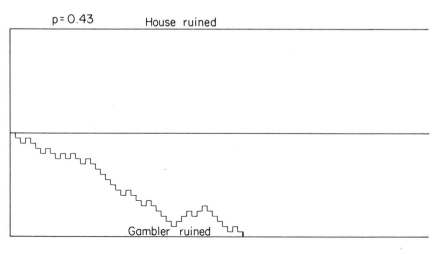

Figure 4.1   Random Walks with Absorbing Barriers

that the gambler will ultimately be ruined if his fortune
before the current play is x. F is the event that the gambler wins the current play. Then evidently $P[E] = q_x$,
$P[E|F] = q_{x+1}$, $P[F] = p$, etc., and with this association,
(4.1) is an application of (4.2).

Equation (4.1) is termed a difference equation. A
book that explains an operational calculus for difference
equations is (E. I. Jury, 1964), Theory and Application of
the Z-Transform Method. Other books on difference equations
include (Freeman, 1965) and (Goldberg, 1958).

A difference equation is a relationship defined on
sequences $\{g(n)\}$. A relationship such as (4.1) of the form

$$\sum_{j=0}^{N} a_j g(n + j) = 0, \ a_N \neq 0,$$

is a homogeneous, time-invariant, linear difference equation
of order N. One may verify that, as in linear differential
equation theory, the set of solutions to the homogeneous
linear equation comprises a vector space of dimension N
over the field of complex numbers. The reader may check to
see that $g(x) = 1$ and $h(x) = (q/p)^x$ are both solutions to
the homogeneous equation (4.1). In the manner of standard
methods of differential equation theory, we set $q_x = Ag(x)$
$+ Bh(x)$ and solve for the constants A and B so that the
boundary conditions $q_0 = 1$, and $q_a = 0$ are satisfied. This
gives

$$q_x = \frac{(q/p)^a - (q/p)^x}{(q/p)^a - 1} , \tag{4.3}$$

a solution which is valid when $p \neq q$. If $p = q$, observe
that $k(x) = x$ is now a solution to (4.1) to conclude, after
satisfying boundary conditions, that

$$q_x = 1 - x/a. \hspace{4cm} (4.4)$$

For instance, if the gambler has \$10, the house has \$20, and
the game is fair, (4.4) tells us that the probability that
the gambler will be ruined is $1 - 10/30 = 2/3$. For playing
red on the standard roulette wheel having two green numbers,
$p = 18/38 \approx 0.474$. If the gambler and the casino again
have \$10 and \$20 respectively, from (4.3) the gambler's
probability of ruin is

$$q_{10} = [(0.526/0.474)^{30} - (0.526/0.474)^{10}]$$

$$\div [(0.526/0.474)^{30} - 1] \approx 0.92.$$

Thus, in changing p from .5 to .474, the probability
of breaking the bank changed from 0.33 to .08, that is, the
probability of avoiding ruin fell to less than 1/4 of the
previous value.

Program 4.1 simulates the roulette situation just
described, printing out the relative frequency that the
gambler broke the bank as the number of repetitions of the
gambling process increases. From the preceding discussion,
the theoretical probability of the bank being "ruined" is
0.08.

In Figure 4.2, we have provided graphs of the probabilities of ruin as computed by means of (4.3) and (4.4).
Each of the subfigures is associated with a different

amount a of money in circulation (the sum of the total
fortunes of the bank and the gambler). Each curve in the
individual subfigures is associated with a certain ratio of
gambler's fortune to the money in circulation. The gambler's
initial fortune is denoted by x. Each of these curves give
the probability of ruin as a function of the probability p
of winning a play. Notice in particular how sharply the
probability of ruin shifts as p varies in the neighborhood
of 1/2. This is especially true for relatively large values
of a.

The book (Casti and Kalaba, 1973) derives the solution
of the gambler's ruin problem as illustration of the notion
of "invariant imbedding," an increasingly popular technique
for computation of solutions of differential and integral
equations.

Program 4.1

Simulates Gambler's Ruin in Roulette

```
      PROGRAM ROULTT (OUTPUT,TAPE6=OUTPUT)
C       ROULTT PLAYS ROULETTE GAMES TO RUIN AND FINDS RELATIVE
C       FREQUENCIES THAT THE HOUSE IS RUINED.

C       HSEFOR- HOUSE'S CURRENT FORTUNE
C       GAMFOR- GAMBLER'S CURRENT FORTUNE
C       IGAME- NUMBER OF GAMES PLAYED TO RUIN
C       NGAM- NUMBER OF TIMES HOUSE IS RUINED
C       RGAM- RELATIVE FREQUENCY HOUSE IS RUINED
C       PLAY- PLAYS ROULETTE FOR ONE TOSS

      INTEGER HSEFOR, GAMFOR
      DATA NGAM /0/

      WRITE(6,80)
80    FORMAT (1H1,30X,"NUMBER      REL FREQ"/
     1         31X,"OF GAMES    BANK RUINED"//)

      DO 100 IGAME=1,1000
C       PLAY TO RUIN WITH BETS OF 1
        GAMFOR = 10
        HSEFOR = 20
200     CALL PLAY(1,GAMFOR,HSEFOR)
        IF( GAMFOR .GT. 0  .AND.  HSEFOR .GT. 0 ) GOTO 200
        IF( GAMFOR .GT. 0 ) NGAM = NGAM + 1
```

```
C          CALCULATE AND PRINT REL FREQUENCIES EVERY 50-TH GAME
           IF( MOD( IGAME, 50) .NE. 0) GOTO 100
           RGAM = FLOAT(NGAM)/FLOAT(IGAME)
           WRITE(6,81) IGAME, RGAM
81         FORMAT(33X,I4,1X,F12.5)

100        CONTINUE
           STOP
           END


           SUBROUTINE PLAY( BET, GAMFOR, HSEFOR )
C          PLAY PLAYS ONE TOSS OF ROULETTE WITH BET "BET"

C             BET- AMOUNT OF BET
C             GAMFOR- GAMBLER"S FORTUNE
C             HSEFOR- HOUSE'S FORTUNE
C             PRWIN- PROBABILITY THAT GAMBLER WINS (18/38)

           INTEGER BET, GAMFOR, HSEFOR
           DATA PRWIN /.4736842105/

           IF( RANF(DUMMY) .LE. PRWIN ) GOTO 100
           HSEFOR = HSEFOR + BET
           GAMFOR = GAMFOR - BET
           RETURN
100        CONTINUE
           HSEFOR = HSEFOR - BET
           GAMFOR = GAMFOR + BET
           RETURN
           END
```

NUMBER OF GAMES	REL FREQ BANK RUINED
50	.08000
100	.06000
150	.07333
200	.07500
250	.08000
300	.08333
350	.08000
400	.08500
450	.08444
500	.07800
550	.07636
600	.07667
650	.07692
700	.07286
750	.07600
800	.07875
850	.08235
900	.08000
950	.08211
1000	.08200

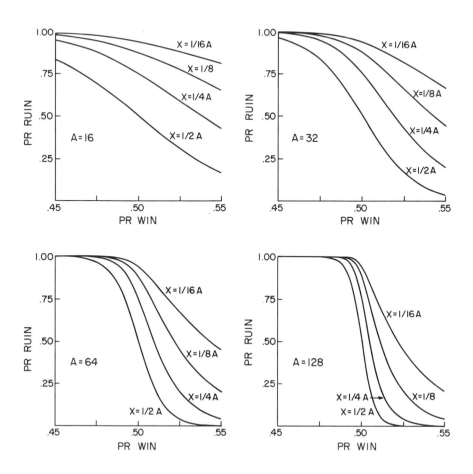

Figure 4.2   Probability of Ruin for Various A, X, and P

4.2   THE CONSERVATION OF FAIRNESS AND THE FUTILITY OF

      GAMBLING SCHEMES

As pointed out in the introductory section, the principle of fairness implies that, generally speaking in a fair
casino,* no matter what the initial fortunes of the gambler
and the casino and no matter what strategy the gambler
follows, his expected return is exactly zero.   In particular, in the gaming situation described in Section 4.1, if
$p = 1/2$, the expected gain associated with the strategy
"Quit when you are \$N ahead," is 0.   To verify the conser-
vation principle in this case, observe that the gambler's
decision to quit when he is ahead by \$N is equivalent to
the game that results if the casino's initial capital is
\$N, the gambler's is \$X, and the gambler plays until either
he or the house is ruined.   The expected return E[R] of a
play of this game is

   E[R] = (-\$X)P[Gambler is ruined]

        + (\$N)P[Gambler breaks the bank].   (4.5)

By (4.4), the respective probabilities are

   P[Gambler is ruined] = q_X = 1 - (X/(N + X)),

   P[Gambler breaks the bank] = 1 - q_X = X/(N + X) . (4.6)

---

*A casino in which all individual plays (or gambles) have
expected return of zero.

Substituting these probabilities into (4.5), we find

$$E[R] = (-\$X)(1 - X/(N + X)) + (\$N)(X/(N + X))$$

$$= (N + X)^{-1}(-XN - X^2 + X^2 + NX) = 0. \qquad (4.7)$$

Let us ask about more general gambling schemes. In particular, the analytic gambler wants to know, precisely, "Is there any way I can play so that I can expect to come out ahead?" We assume the options available are "continue playing" and "stop and go home." In the section to follow, we study games which allow the gambler to vary the stakes from play to play.

An elementary construct of decision theory (see, for example, Optimal Statistical Decisions (DeGroot, 1970)) is used. Let $\{X_i\}$ denote a random walk (i.e., a history of fortunes).

Definition. Assume a nonterminating random walk $\{X_i\}$ has been specified. A stopping time is a binary-valued function $t(x_1, x_2, \ldots, x_n)$ defined on initial segments $(x_1, \ldots, x_n)$ of observations of walks. The function t is assumed to have the property that stopping is sure to eventually occur, i.e.,

$$P[t(X_1, \ldots, X_n) = 0 \text{ for all } n] = 0.$$

The interpretation of "stopping time" is that the gambler quits the first time n that $t(x_1, \ldots, x_n) = 1$ and goes home (with amount $x_{t(x_1, \ldots, x_n)}$). He continues gambling if $t(x_1, \ldots, x_n) = 0$. Of course, t = 1 if ever the gambler or the house is out of funds. The conservation of fairness

<u>principle</u> implies that for fair games

Expected Terminal Fortune = $E[X_t] = x_1$

= Initial Fortune,

for all stopping times t. Later in this section we will discuss conditions under which this principle can be shown to hold, and circumstances in which it is violated.

We next consider a stopping time strategy. Let us generalize the earlier random walks by allowing the step size to be other than 1 and -1. If fact, our randomizing device for the computer experiment to follow will be a die. The payoff at each play is $2k, where k is the number of dots on the top face. The expected payoff is

$$\$E[K] = \$2(1 + 2 +\cdots+ 6) \times 1/6 = \$7. \tag{4.8}$$

The casino is assumed to charge $7 for each toss, and thus the gamble is fair. In hope of defying the "conservation of fairness" we will adopt a complicated stopping time. Let k_j be the number of dots showing on the <u>jth</u> toss. N is defined by

$$N = \sum_{j=1}^{v} k_j$$

where v is defined to be the first toss j such that k_j is even. Finally

$$S_N = \sum_{j=1}^{N} k_j.$$

The stopping time is defined by the rule: stop at time S_N (or before if you are out of funds or have broken the bank). We will assume that initially the house has \$200 and the gambler has \$100.

With some difficulty we could calculate the expected worth of the game if the above stopping rule is employed, but instead, in the spirit of the Monte Carlo method, we will simply experiment by playing a large number of games using this rule. The conservation of fairness will be confirmed in this case if the average return in a large number of plays tends to the initial fortune, which in this case is 100. Program 4.2 below performs this experiment.

Program 4.2

Simulates a Stopping Strategy for Dice Gambles

```
      PROGRAM CONFAR (OUTPUT,TAPE6=OUTPUT)
C        CONFAR DEMONSTRATES THE CONSERVATION OF FAIRNESS
C          PRINCIPLE BY SIMULATING A GAME WITH A COMPLICATED
C          STOPPING RULE MANY TIMES.

C        HSEFOR- HOUSE'S FORTUNE
C        GAMFOR- GAMBLERS FORTUNE
C        N,VEVEN- USED TO CALCULATE THE STOPPING TIME
C        SN- THE STOPPING TIME
C        GAMTOT- TOTAL RETURNS TO THE GAMBLER
C        AVERET- AVERAGE RETURNS TO THE GAMBLER
      LOGICAL VEVEN
      INTEGER SN, GAMFOR, HSEFOR, GAMTOT

      WRITE (6,80)
80    FORMAT(1H1,10X,"NUMBER      AVERAGE"/
     1            11X,"OF GAMES   RETURN"/
     2            11X,"PLAYED"/)
      DO 100 IGAME=1,10000
        HSEFOR = 200
        GAMFOR = 100
        N = 0
        VEVEN = .FALSE.
        SN = 0
        J = 0
```

```
200        J = J + 1
           KJ = 1.0 + 6.0*RANF(DUMMY)
           IF( .NOT. VEVEN ) N = N + KJ
           IF( MOD(KJ,2) .EQ. 0 ) VEVEN = .TRUE.
           IF( J .LE. N ) SN = SN + KJ

           GAMFOR = GAMFOR + 2*KJ - 7
           HSEFOR = HSEFOR - (2*KJ - 7)

           IF(.NOT.(HSEFOR.LT.14.OR.GAMFOR.LT.7.OR.J.EQ.SN) ) GOTO 200
           GAMTOT = GAMTOT + GAMFOR
           HSETOT = HSETOT + HSEFOR

           IF( MOD(IGAME,1000) .NE. 0 ) GOTO 100
              AVERET = FLOAT(GAMTOT)/FLOAT(IGAME)
              WRITE(6,81) IGAME, AVERET
81            FORMAT(9X,I6,5X,G12.5)

100        CONTINUE
           STOP
           END
```

NUMBER OF GAMES PLAYED	AVERAGE RETURN
1000	100.17
2000	99.969
3000	99.836
4000	100.03
5000	100.12
6000	100.06
7000	100.06
8000	100.01
9000	100.05
10000	100.16

Feller, p. 198, sets forth the motivation for the principle of conservation of fairness in asserting:

> The painful experience of many gamblers has taught us the lesson that no system of betting is successful in improving the gambler's chances. If the theory of probability is true to life, this experience must correspond to a provable statement.

The mathematics behind this "provable statement" is too advanced for this book. But we refer the reader to a generalization of Wald's lemma given by Theorem 2.3 in the scholarly book Great Expectations: The Theory of Optimal Stopping (Chow et al., 1971). This theorem implies that if $\{X_n\}$ is a bounded random sequence (in our case, bounded by the capital of gambler and the bank) and if the individual

gambles are fair (i.e., $E[X_n|X_{n-1}] = X_{n-1}$), then for all
stopping rules t, the conservation of fairness principle
holds:

Expected Terminal Fortune = $E[X_t] = x_1$ = Initial Fortune.

If the sizes of the possible fortunes are not bounded, the
situation is vastly changed. There is no bound to how much
money the gambler can make. For the random walk is certain
to eventually exceed any number M. To show this for repeated tosses of a fair coin, we appeal to the "law of the
iterated logarithm" (described, for example, in Feller,
Section VIII 5), which, when stated in our notation, assures
us that for any positive $\lambda < 1$, with probability 1,

$$x(n) > \lambda\sqrt{\log(\log(n))}$$

for infinitely many n. So to earn \$M with probability 1,
the gambler can adopt the rule "stop if and only if $x_n \geq M$."
Another curiosity is that no matter what rule the gambler
adopts, there is always one that results in higher expected
payoff. But, of course, the assumption of unlimited capital
is entirely unrealistic, and if you assume the capital is
limited--no matter how large--then the game is fair, by
Wald's lemma.

4.3   GAMBLING IN SUBFAIR CASINOS

Whereas in earlier sections of this chapter, the gambler has been restricted to a fixed bet size (and his only
options are to bet again or go home), in this section, he
is allowed a more diverse strategy for the coin toss game:
He may bet any amount he wants to, with the proviso that he
cannot bet what he does not have; the amount he bets on a
coin toss must be no greater than his present total fortune.

Our opportunity of variable bet sizes vastly compli-
cates the mathematics of the problem.   In particular, the
gambler's ruin problem becomes much more difficult to solve.
But a solution to the gambler's ruin problem, in the case
of subfair coin-toss games (the coin is biased against the
gambler), is available in the guise of the problem of a
gambler who desires to attain some fixed fortune F larger
than his initial fortune, and he therefore seeks a strategy
to maximize the probability of achieving this fortune.
Since the gamble is sub-fair, the probability of attaining
F cannot be larger than $x/(F + x)$, where x is his initial
fortune.   How large can he make it?   Dubins and Savage
(1965, p. 1) describe this problem picturesquely:

> Imagine yourself at a casino with $1,000.   For
> some reason, you desperately need $10,000 by
> morning; anything less is worth nothing for
> your purposes.   What ought you do?   The only
> thing possible is to gamble away your last cent,
> if need be, in an attempt to reach the target
> sum of $10,000.   There may be a moment of
> moral confusion and discouragement.   For who
> has not been taught how wrong and futile it is
> to gamble, especially when short of funds?
> Yet, gamble you must or forgo all chance of
> the great purpose that can be achieved only
> at the price of $10,000 payable at dawn.   The
> question is how to play, not whether.

We may discuss this problem in the gambler's ruin
setting: Just take $9,000 as the house's fortune. We
allow, as strategy σ, any algorithm for determining how
much to bet at play n, on the basis of the entire past
history of the game. It is natural to define the probability of ruin as

$$\text{minimum}_\sigma \; p(\sigma),$$

where $p(\sigma)$ denotes the probability of ruin under strategy σ.

Intuitively, it is a bad policy to bet small amounts
at each play, for the law of large numbers (Section 3.1)
implies that the average return will converge to the expectation per play, which is negative (recall, we are playing
with a subfair coin, or, in the description in Dubins and
Savage, playing red and black in roulette, which is subfair
because of the green partitions on the wheel). Specifically,
let p denote the probability of the gambler's winning a coin
toss, and suppose he bets some fixed sum s at each play for
a large number N of plays. Let W_j denote the return at the
jth play. Thus $\mu_W = E[W_j] = s(2p - 1)$ and σ_W^2 = variance of
$W_j = 4s^2 p(1 - p)$. X_n, the net change in fortune at time n,
is then expressible as $X_n = \sum_{i=1}^{n} W_i$. The central limit
theorem, (Lindgren, p. 143) which was briefly discussed in
Section 2.3, in connection with a particular normal variate
generator, asserts that if $\{W_i\}$ is a sequence of independent, identically-distributed random variables with mean
μ_W and variance σ_W^2 and if $X = \sum_{i=1}^{n} W$, then

$$\frac{X_n - n\mu_W}{(n\sigma_W)^{1/2}} \to Y,$$

in distribution, where Y is the standard normal variate.

Thus for large (or even moderate) n,

$$P[X_n \leq 0] \simeq P[Y \leq \sqrt{n}\mu_W/\sigma_W]$$

$$= P[Y \leq \sqrt{n}(2p - 1)/\sqrt{4p(1 - p)}],$$

and as n increases, the probability defined on right-hand side converges to 1. Thus for large n, the gambler is almost certain to be "in the hole." And yet if the bet size s is small, n must necessarily become large.

In summary, if the gambler makes small enough bets to allow the central limit theorem to apply, his situation is hopeless. The opposite of "small and many bets" is a bold strategy: bet all you have, or, if you have over $5,000, bet the amount you need so that if you win, you get exactly the goal amount $10,000. In Theorem 1, Chapter 5 of Dubins and Savage (1965) it is seen that playing boldly is indeed an optimal strategy.

In Program 4.3, we experimentally compare the bold strategy with the strategies of betting $100 at each toss and $1,000 at each toss. The program performs a large number of iterations of coin toss under each strategy and compares the relative frequency of ruin.

Program 4.3

Compares Three Betting Strategies For Roulette

```
           PROGRAM SUBFAR (OUTPUT,TAPE6=OUTPUT)
C             SUBFAR COMPARES THE BOLD STRATEGY WITH THE BET $1000 AND
C             BET $100 EACH PLAY STRATEGIES IN A SUBFAIR GAME (ROULETTE)
C             BY SIMULATING 10000 GAMES EACH PLAYED TC RUIN FOR
C             EACH STRATEGY.

C                GAFOR- GAMBLER'S FORTUNE
C                HSEFOR- HOUSE'S FORTUNE
C                NBOLD, NTHOU, NHNRD- NO OF TIMES HOUSE IS RUINED
C                    FOR EACH STRATEGY
C                RBOLD,RTHOU,RHNRD- RELATIVE FREQUENCY OF RUIN
C                    FOR EACH STRATEGY
C                PLAY- PLAYS ROULETTE FOR ONE TOSS.  SEE PROGRAM 4.1

           INTEGER GAMFOR, HSEFOR
           DATA NBOLD, NTHOU, NHNRD /3*0/

           WRITE(6,80)
80         FORMAT(1H1,25X,"RELATIVE FREQUENCY THAT"/
          1          26X,"  PLAYER WINS $10000"//
          2          21X,"NUM OF  BOLD      $1000      $100"/
          3          21X,"GAMES             BETS       BETS"//)

           DO 200 IGAME=1,10000
C             PLAY BOLDLY TO RUIN- BET AS MUCH AS NEEDED OR POSSIBLE
           GAMFOR = 1000
           HSEFOR = 9000
100        CALL PLAY( MINO(GAMFOR, 10000-GAMFOR), GAMFOR, HSEFOR )
           IF( GAMFOR .GT. 0 .AND. HSEFOR .GT. 0 ) GOTO 100
           IF( GAMFOR .GT. 0 ) NBOLD = NBOLD + 1

C             PLAY WITH $1000 BETS TO RUIN
           GAMFOR = 1000
           HSEFOR = 9000
120        CALL PLAY( 1000,GAMFOR,HSEFOR)
           IF( GAMFOR .GT. 0 .AND. HSEFOR .GT. 0 ) GOTO 120
           IF( GAMFOR .GT. 0 ) NTHOU = NTHOU + 1

C             PLAY WITH $100 BETS TO RUIN
           GAMFOR = 1000
           HSEFOR = 9000
140        CALL PLAY( 100, GAMFOR, HSEFOR )
           IF( GAMFOR .GT. 0 .AND. HSEFOR .GT. 0 ) GOTO 140
           IF( GAMFOR .GT. 0 ) NHNRD = NHNRD + 1

C             CALCULATE RELATIVE FREQUENCIES EVERY 500-TH GAME
           IF( MOD(IGAME,500) .NE. 0 ) GOTO 200
             RBOLD = FLOAT(NBOLD)/FLOAT(IGAME)
             RTHOU = FLOAT(NTHOU)/FLOAT(IGAME)
             RHNRD = FLOAT(NHNRD)/FLOAT(IGAME)
             WRITE(6,81) IGAME, RBOLD, RTHOU, RHNRD
81           FORMAT (21X,I5,3G10.3)

200        CONTINUE

           STOP
           END
```

```
          RELATIVE FREQUENCY THAT
          PLAYER WINS $10000

   NUM OF    BOLD       $1000      $100
   GAMES                BETS       BETS


     500    .840E-01   .540E-01   0.
    1000    .810E-01   .630E-01   0.
    1500    .913E-01   .627E-01   .667E-03
    2000    .890E-01   .595E-01   .500E-03
    2500    .868E-01   .612E-01   .400E-03
    3000    .837E-01   .607E-01   .333E-03
    3500    .831E-01   .591E-01   .286E-03
    4000    .843E-01   .583E-01   .250E-03
    4500    .851E-01   .609E-01   .222E-03
    5000    .844E-01   .594E-01   .200E-03
    5500    .860E-01   .605E-01   .182E-03
    6000    .858E-01   .627E-01   .167E-03
    6500    .845E-01   .634E-01   .154E-03
    7000    .836E-01   .631E-01   .143E-03
    7500    .843E-01   .632E-01   .133E-03
    8000    .865E-01   .633E-01   .125E-03
    8500    .858E-01   .638E-01   .118E-03
    9000    .860E-01   .633E-01   .111E-03
    9500    .846E-01   .622E-01   .105E-03
   10000    .847E-01   .616E-01   .100E-03
```

4.4   RUIN OF AN INSURANCE COMPANY

The situation of an insurance company doing business
is akin to that of a gambler playing in an infinitely
wealthy casino: There is only one absorbing barrier,
namely financial collapse resulting from more claims than
assets.   For the purposes of keeping this discussion elementary, we make some vastly naïve assumptions: 1) The number N of policies is constant, 2) The overhead amount H
(not including the claims) is the same each year, and
3) An amount proportional to the total capital is removed
from the assets to pay dividends each year.   The constant
of proportionality is d.   P denotes the annual policy premium, the random variable C_j denotes the cash amount of
claims in year j, and k_j denotes the capital at the end of
the year j.   Under normal circumstances, the capital obeys
the (stochastic) linear difference equation

$$k_j = (1 - d)(k_{j-1} - C_j + NP - H). \tag{4.9}$$

Under extraordinary circumstances, the company's capital
is wiped out by claims, and it goes out of business. This
happens at any time j such that $C_j + H > k_{j-1} + NP$, assuming
cash flow takes place at one time in a given year. The
reader will be able to view this process, then, as a random
walk with a lower absorbing barrier. The nature of the walk
determined by this insurance process is more complicated
than the gambler's ruin problem because the step size is
not assumed to be plus or minus 1.

   We will illustrate the employment of the Monte Carlo
method in operations research by studying the decision
problem engendered when one supposes that the insurance
company can, for an annual fee F, become underwritten by a
bigger company. Underwriting provides, in our problem,
that should the claims ever exceed the assets of the company,
a certain amount of additional money, S, will be made
available. From this point on, the assurance policy is
cancelled. This has the effect of lowering the absorbing
barrier by S units. We assume the company's goal is to
maximize total dividend payments over an M-year operating
horizon. One must decide whether the company should buy
the underwriting policy.

   We will now use simulation to solve such a decision
problem. It will be assumed that C_j is distributed as an
exponential variable with parameter 10. The horizon M is
set to 50 years; the initial capital k_0 is 50, the underwriting policy is for S = 20, and F, the annual cost of the
underwriting policy, is 1.25. H = 4 and d = 0.1. NP, the
annual premium income, is 15.

Program 4.4 repeatedly runs walks according to the recursive rule (4.12) and the parameter values as described above under regimes with and without underwriting. It computes the cumulative dividends and stops whenever the average (over the simulation iterations) dividends, with underwriting and otherwise, differ between any two iterations by an amount less than 1/10 the difference of the average dividends. That is, if A_k denotes the average dividends under assurance computed on the basis of k simulation runs and B_k is the average of the dividends when no assurance policy is purchased, simulation terminates whenever

$$\max \left\{ |A_k - A_{k-1}|, \ |B_k - B_{k-1}| \right\} < 1/10 |A_k - B_k| .$$

The law of large numbers implies that termination is certain to occur eventually provided the expected return under the two possible plans differ. The reader is advised that there is a positive probability, under this scheme, that a wrong decision will be made. He must hope that this probability of error is small. Through additional study, one could develop a rule for bounding and limiting this probability of decision error.

In Freiberger and Grenander (1971) and Grenander (1973), more elaborate insurance problems are discussed and studied by the Monte Carlo method. The use of simulation for solving non-trivial stochastic optimization problems is popular. Hammersley and Handscomb (1964) assert

> Monte Carlo methods tend to flourish on problems that involve a mass of practical complications of the sort encountered more and more frequently as applied mathematics and operational research come to grips with actualities.

Program 4.4

Simulates Two Insurance Regimes

```
      PROGRAM INSUR(OUTPUT,TAPE6=OUTPUT)
C         INSUR SIMULATES OPERATION OF AN INSURANCE COMPANY
C         UNDER TWO DIFFERENT POLICIES FOR MANY M YEAR
C         OPERATING HORIZONS.  EXPECED DIVIDENDS UNDER EACH
C         POLICY ARE COMPARED.

C         KO,KT- INITIAL CAPITAL, CAPITAL AT END OF YEAR T
C         M- OPERATING HORIZON IN YEARS
C         S- AMOUNT OF ASSURANCE
C         F- COST OF ASSURANCE/YEAR
C         H- OVERHEAD/YEAR
C         D- DIVIDEND RATE
C         NP- TOTAL PREMIUM INCOME/YEAR
C         AK,AKM1- SUCCESSIVE AVEAGE TOTAL DIVIDENDS WITHOUT
C         BK,BKM1-  AND WITH ASSURANCE
C         C- ANNUAL CLAIMS, EXPONENTIALLY DISTBIBUTED
C         ADIV,BDIV- DIVIDENDS WITH AND WITHOUT ASSURANCE

      REAL KO,KT,NP
      DATA KO,M,S,F,H,D,NP /50.C, 50, 20.0, 1.25, 4.0, 0.1, 15.0 /

      K = 0
      AK = 0.0
      BK = 0.0

100   CONTINUE
      K = K + 1
      AKM1 = AK
      BKM1 = BK
C         SIMULATE WITHOUT ASSURANCE
      ADIV = 0.0
      KT = KO
      DO 120 I=1,50
        C = -10.0*ALOG( RANF(DUMMY) )
        IF( KT - C + NP - H .LT. 0.0 ) GOTO 125
        ADIV = ADIV + D*( KT - C + NP - H )
        KT = (1.0 - D)*( KT - C + NP - H )
120   CONTINUE
125   AK = ( FLOAT(K-1)*AKM1 + ADIV )/FLOAT(K)



C         SIMULATE WITH ASSURANCE
      BDIV = 0.0
      FT = F
      KT = KO
      DO 140 I=1,50
        C = -10.0*ALOG( RANF(DUMMY) )
        IF( KT - C + NP - H - FT .GT. 0.0 ) GOTO 142
C         ASSURANCE PAYS OFF AND IS CANCELED
          KT = KT + S
          FT = 0.0
          S = 0.0
142     IF( KT - C + NP - H - FT .LT. 0.0 ) GOTO 145
        BDIV = BDIV + D*( KT - C + NP - H - FT )
        KT = (1.0 - D)*( KT - C + NP - H - FT )
140   CONTINUE
145   BK = (FLOAT(K-1)*BKM1 + EDIV )/FLOAT(K)

      IF( AMAX1( ABS(AK - AKM1), ABS(BK - BKM1) ) .GT.
     1       ABS( AK - BK )/10.0 ) GOTO 100
```

```
            WRITE (6,80)  K,  AK,  BK
   80       FORMAT (1H1,10X,"RUIN OF AN INURANCE COMPANY"//
      1              13X,I3," 50 YEAR PERIODS WERE SIMULATED"/
      2              11X,F8.2," IS AVERAGE DIVDEND WITHOUT ASSURANCE"/
      3              11X,F8.2," IS THE AVERAGE DIVIDEND WITH ASSURANCE")
            STOP
            END
```

```
        RUIN OF AN INURANCE COMPANY

        18 50 YEAR PERIODS WERE SIMULATED
        83.89 IS AVERAGE DIVDEND WITHOUT ASSURANCE
        59.25 IS THE AVERAGE DIVIDEND WITH ASSURANCE
```

4.5 GAMBLER'S RUIN APPLICATIONS IN INVENTORY CONTROL

Imagine a gambler repeating plays of a subfair game
with a fixed bet size. Suppose at fixed periodic intervals he can request and receive any desired amount of
money, and suppose further he is regularly taxed for a proportion of the amount of money in his possession, but on
the other hand, he suffers a penalty if he runs out of money.
A final factor in this model is that the time it takes for
requested money to be delivered may be random. The gambler's
wish is only to continue the gambling process in as inex-
pensive a way as possible.

It must be admitted that this is an unlikely gambling
situation, but mathematically, this model is a restatement
of the popular and important periodic review inventory control problem (e.g. Hadley and Whitin, 1963, Chapter 5) which
is: Given a variable D representing the daily demand for
goods of a certain type, given a daily storage charge (known
as a holding cost) i for each item in stock, and given a
certain backorder penalty π for orders taken when desired
goods are not in stock, find a periodic ordering policy

which minimizes the expected annual cost of doing business.
The variable D plays the role of the gamble outcome and the
holding cost is akin to the taxation in our artificial gambling problem.

Let T be the review period; that is, orders may be
placed only on days which are integral multiples of T. The
lead time (denoted by τ) is defined to be the time which
elapses between the placement of an order and its delivery.
It is known from inventory analysis (Hadley and Whitin,
1963, for example) that for problems of this type, an optimal policy is to order R* - x(t) items, x(t) denoting the
current inventory level and R* being a certain number which
depends on the parameters of the inventory problem.   This
policy is called an order up to R strategy and in the words
of Hadley and Whitin (1963, p. 237) is "the most widely used
doctrine for periodic review systems."

For fixed review period T and constant lead time τ, it
is well known (Johnson and Montgomery, 1974, Section 2-6.3,
for example) that the optimum ordering quantity R* is approximated by the solution to the equation

$$\int_{R^*}^{\infty} f(x) \, dx = iT/\pi, \qquad\qquad (4.10)$$

where f is the probability density for the cumulative demand
from time 1 to T + τ. The approximation (4.10) is accurate
only when the cost π of backorders is much larger than the
holding cost, i.

In our first Monte Carlo study we chose the parameters
of a periodic review system considered by Johnson and
Montgomery (1974, p. 58). For this system the holding cost

i is 0.4¢/item-day, the backorder cost π is $30 per item
backordered, the ordering and review cost is $15, T is 75
days, and τ is 62 days. The yearly demand is approximately
normal with mean 500 and variance 800. Take a year to be
250 working days. Then a daily demand which gives this
yearly demand is $D = \sum_{i=1}^{N} G_i$ where N has a Poisson distribution with parameter 20/13 and each G_i has a geometric
distribution with parameter 10/13. N can be thought of as
the number of orders recieved on a given day and G_i as the
size of the ith order. The variable D is known in the inventory control literature as a stuttering Poisson process
(Hadley and Whitin, 1963, p. 113).

For different values of R we thrice simulated runs of
100 review periods each. Figure 4.3 compares sample mean
annual costs with its theoretical approximation obtained
from inventory theory in Johnson and Montgomery (1974, p. 59).

Our next Monte Carlo study allowed for random lead
time τ. Basic to calculation of optimal ordering level R*
is the ability to calculate the expected number of backorders. Hadley and Whiten (1963, p. 287) assert that this
quantity may be found analytically only for very special
forms of lead time distribution. In Program 4.5, τ is taken
to be uniformly distributed on the interval [2T/3, T]. A
plot (Figure 4.4) of sample annual average cost vs. level
R, as in Figure 4.3, is provided for the random lead time
case, suggesting that the optimal ordering level R* may be
closely approximated by examining the graph obtained from
this simulation study.

One may continue this line of attack on the inventory
problem to simultaneously approximate the optimal review
period T as well as the optimal level R*(T). The idea here

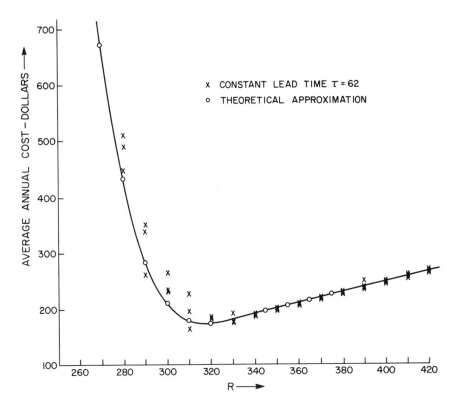

Figure 4.3   Simulated Cost of a Fixed Lead-Time Inventory
            Process

Program 4.5

Simulates an Inventory Process with Random Lead Time

```
      INTEGER FUNCTION RDEMD(DUMMY)
C         RDEMD RETURNS A RANDOM DAILY DEMAND WITH A COMPOUND
C         POISSON DISTRIBUTION.

C         EXPLAM- EXP(-LAMBDA) WHERE LAMBDA IS THE PARAMATER
C         OF THE POISSON DISTRIBUTION
C         ALG1MP- ALOG(1.0 - P) WHERE P IS THE PARAMATER OF
C         THE GEOMETRIC DISTRIBUTION

C         THESE VALUES GIVE A MEAN OF 2.0 AND VARIANCE OF 3.2
      DATA EXPLAM, ALG1MP /0.2147111724, -1.466337069 /

C         GENERATE N WITH A POISSON DISTRIBUTION
      N = 0
      S = 1.0
100   S = S*RANF(DUMMY)
      IF( S .LT. EXPLAM ) GOTO 200
      N = N + 1
      GOTO 100

C         SUM N GEOMETRIC RANDOM VARIABLES
200   RDEMD = 0
      IF( N .EQ. 0 ) RETURN
      DO 300 I = 1,N
      RDEMD = RDEMD + INT( ALOG(RANF(DUMMY))/ALG1MP) + 1
300   CONTINUE
      RETURN
      END
      PROGRAM INVTRY (OUTPUT,TAPE6=OUTPUT)
C         INVTRY SIMULATES AN INVENTORY SYSTEM WITH RANDOM LEAD
C         TIMES AND FINDS THE AVERAGE YEARLY COST.

C         P- ORDER UP TO THIS LEVEL
C         T- REVIEW PERIOD IN DAYS
C         TAU- LEAD TIME IN DAYS- A RANDOM VARIABLE
C         HLDCST- HOLDING COST/ITEM/DAY
C         REVCST- REVIEW AND ORDER COST/PERIOD
C         BAKCST- BACK ORDER COST/ITEM
C         INV- NUMBER OF ITEMS ON HAND AT END OF A DAY
C         ORD- NUMBER OF ITEMS ORDERED LAST REVIEW PERIOD
C         D- DAILY DEMAND- A RANDOM VARIABLE
C         DMEAN- MEAN OF D
C         RDEMD( )- DEFINES RANDOM DAILY DEMAND
C         RLEAD( )- DEFINES RANDOM LEAD TIME
C         TOTCST- TOTAL COST FOR ALL PERIODS
C         YRCST- YEARLY COST
C         N- NUMBER OF REVIEW PERIODS SIMULATED

      INTEGER R, T, TAU, INV, ORD, D, RDEMD, RLEAD
      DATA HLDCST, REVCST, BAKCST /0.004, 15.00, 30.00 /
      DATA DMEAN, T, N /2.0, 75, 100 /

      RLEAD(DUMMY) = 51 + INT( 25.0*RANF(DUMMY) )
```

```
          WRITE(6,80) N, T
80        FORMAT(1H1,"SIMULATION OF AN INVENTORY SYSTEM"/
      1            1X,"FOR",I4," REVIEW PERIODS OF LENGTH",I3," DAYS"///
      2            1X,"ORDER    AVERAGE"/
      3            1X,"UP TO    ANNUAL"/
      4            1X," R       COST"//)

          DO 300 R=250,400,10
C           INITIALIZE INVENTORY AND ORDER
            INV = DMEAN*FLOAT(T)
            ORD = MAX0( R - INV, 0 )
            TAU = RLEAD(DUMMY)
            TOTCST = 0.0
            DO 200 I=1,N
C             SIMULATE FOR ONE REVIEW PERIOD
              DO 100 K=1,T
C               DID ORDER ARRIVE?
                IF( K .EQ. TAU )  INV = INV + ORD
                D = RDEMD(DUMMY)
C               BACKORDERS?
                  IF(INV - D .LT. 0) TOTCST = TOTCST + BAKCST*(D - INV)
                  INV = MAX0(INV - D, 0)
                  TOTCST = TOTCST + FLOAT(INV)* HLDCST
100             CONTINUE
C             REVIEW
              ORD = MAX0( R - INV, 0 )
              TAU = RLEAD(DUMMY)
              TOTCST = TOTCST + REVCST
200           CONTINUE
            YRCST = TOTCST/FLOAT(T*N)*250.0
            WRITE(6,81) R, YRCST
81          FORMAT(1X,I4,4X,G10.4)
300         CONTINUE

          STOP
          END
```

```
              SIMULATION OF AN INVENTORY SYSTEM
              FOR 100 REVIEW PERIODS OF LENGTH 75 DAYS

              ORDER    AVERAGE
              UP TO    ANNUAL
                R      COST

               250     1689.
               260     1349.
               270     895.6
               280     576.2
               290     419.9
               300     299.1
               310     210.0
               320     250.2
               330     194.4
               340     205.1
               350     198.4
               360     204.7
               370     220.9
               380     228.4
               390     242.5
               400     254.6
```

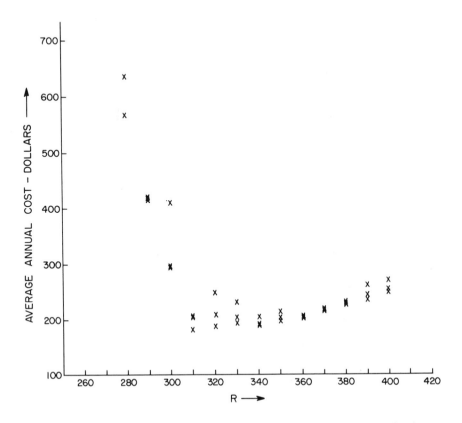

Figure 4.4   Simulated Annual Costs of a Variable Lead-Time
             Inventory Process

is that simulation is an effective means for solving more
realistic inventory models than the textbook type.

In Chapter 1 of their book, Hadley and Whitin (1963)
discuss the weakness of existing inventory system theory in
comparison to the realistic multiechelon processes one would
like to study. They assert that,

> ... perhaps the most important reason [for the
> weakness of existing theory] is that it is
> very difficult to study analytically multi-
> echelon systems.... In fact, very little

work has been done in this area. It will be
seen that even the relatively simple structure
that will be studied can be very complex to
analyze.

As we have tried to suggest in the preceding examples
as well as problems 6 and 7, through simulation one has
effective means of approximating the solution to analytically intractible inventory problems.

4.6 GAMBLER'S RUIN PROCESSES IN STATISTICS

We have seen that the theory of gambler's ruin, the
central model of the discourse of this chapter, has rele-
vancy to the theory of insurance marketing as well as inventory control under random demand. One may see that the inventory control model is applicable to organizations providing services to stochastic numbers of clients, random
decreases of inventories, telephone switchboards servicing
simultaneously a random number of inputs, and so forth.

Feller (p. 253) asserts "The statistician who applies
statistical tests is engaged in a dignified sort of gambling." The statistical technique called the "sequential
probability ratio test" (SPRT) by Wald is an example of a
gambler's ruin problem. The statistician observes independent samples $\{Z_i\}$ whose common density is _either_ f_1 or
f_2; he is to determine which. The SPRT test consists of
computing at each time n the sum

$$\sum_{i=1}^{n} \log(f_1(Z_i)/f_2(Z_i)).$$

This process continues until the sum exceeds a given number B or is less than another given number A, at which time the testing is stopped and the statistician makes the decision that the underlying density was f_1 if the sum is greater than B or f_2 if it was less than A.* We can identify this structure directly with a gambler's ruin problem. The gambler's fortune at any time n is $\sum_{i \leq n} \log(f_1(Z_i)/f_2(Z_i))$, and the return of any gamble is the random variable

$$\log(f_1(Z)/f_2(Z)),$$

where Z has the density common to the sequence $\{Z_i\}$ described above. If the bank is broken (the bank has initial capital B), f_1 is chosen, and if the gambler loses all his money (initially A) f_2 is chosen. Another connection between random walks and statistics is developed in Feller, Section XIV.8.

4.7   REMARKS ON SIMULATION LANGUAGES

There are computer simulation languages available for easy programming of certain operations research simulation problems. While claims are made that simulation languages are universally effective for Monte Carlo problems, a cursory examination shows that they are motivated by queuing, inventory, and job shop studies. For these types of

---

*The SPRT test is described in DeGroot (1970, p. 360f) and, in more detail, in Chernoff (1972).

problems, simulation languages offer the advantage (in comparison with FORTRAN or other general-purpose languages)
of having operations and variable sets which allow the user
to avoid mildly complicated subroutines and bookkeeping
devices. We refer the reader to the paper by Teichroew
and Lubin (1966) or the book by Gordon (1969) for further
reading on these matters.

1) Section 4.1 employs the theory for time-invariant linear difference equations. The present problem introduces us to time-varying linear homogeneous difference equations, which are generally less convenient, computationally. Suppose that the probability p_x of winning a play is proportional to the gambler's current fortune, x. That is, equation 4.1 is now written

$$q_x = p_x q_{x+1} + (1 - p_x) q_{x-1}. \qquad (4.1')$$

Then the method for finding homogeneous solutions given in Section 4.1 is no longer applicable. A brute-force computer approach is available, and is perhaps the most sensible for this problem. Choose $w_0 = 1$, $w_1 = 0$, $v_0 = 0$, and $v_1 = 1$. Construct sequences $W = \{w_i\}$ and $V = \{v_i\}$ by the following recurseve relation implied by (4.1'),

$$w_{i+1} = p_i^{-1}[w^i - (1 - p_i)w_{i-1}]$$

$$v_{i+1} = p_i^{-1}[v_i - (1 - p_i)v_{i-1}] \qquad (4.2')$$

Then, provided that V is not a scalar multiple of W, it is true and reasonably evident that any solution $\{z_x\}$ of (4.1') can be written as a linear combination (i.e., c and d below are scalars) of W and V, i.e.,

$$z_i = cw_i + dv_i, \quad 0 \le i \le a.$$

In particular, using the condition that $q_0 = 1$, $q_a = 0$, we can solve for c and d and thus find q_i for each i.

a) For $x = 2,3,\ldots,7$, find q_x for the probability sequence

$$p(1) = 2/3, \; p(2) = 1/3, \; p(3) = 1/2, \; p(4) = 2/3,$$

$$p(5) = 1/3, \; p(6) = 2/3, \; p(7) = 1/2, \; (a = 8).$$

b) Check your answer to a) by simulating to find the probability of ruin for various fortunes x, $1 < x < 7$.*

2) As described in the books (Maistrov, 1974) and (Smith, 1959), the type of gambling question which inspired Fermat and Pascal in their earliest mathematical works on probability is the following: Two players have put up a pot of $10, total. They play a sequence of gambles in which one player or the other wins a point. The first to achieve a total of five points wins the pot.

---

*For an invariant imbedding solution to this numerical problem, see Chapter 1 of (Casti and Kalaba, 1973).

Now suppose that due to an emergency, the game must be terminated at a point at which Player A has three points and Player B has four points. Assume further that each player has probability of 1/2 of winning any particular play. How should the pot be divided so as to be "fair" to each player?   Provide an answer.

3)   The notion of a higher (than 2) order difference equation arises when we allow a multiplicity of possible outcomes from a single gamble. Find the difference equation for the probability of the gambler's ruin if the basic gamble depends on the outcome of a toss of a single die in such a fashion that the gambler receives in dollars twice the number of points showing, minus $7. In addition to giving the difference equation, deduce the boundary conditions. There should be k boundary conditions for a kth order difference equation. Note: It is possible to solve this difference equation problem by essentially the same method as given in Section 4.1. However, as an intermediate step, one must find all the roots of a fifth degree polynomial (which can easily be done by standard numercial analysis techniques).

4)   Think up some fair gamble and some complicated playing strategy, in the spirit of Section 4.2. Simulate and tabulate the average return from repetitions of the gambling process.

5)   Let us consider the basic die gamble used in problem 2 above (as well as for the simulation example in Section 4.2). That is, the gambler receives twice the number of points showing on a tossed die minus his $7 entry

fee, at each gamble. Suppose that the die is unfair.
Specifically, the probability of j points showing is
given by

$$P[j] = 1/6 + 1/20, \quad j = 1,2,3$$
$$P[j] = 1/6 - 1/20, \quad j = 4,5,6$$

Assume that the gambler can bet on any integer multiple
s of $7, and thence receive 2s multiplied by the number
of dots showing. Through simulation, experimentally
compare the bold betting strategy (described in Section
(4.2) against more timid strategies for maximizing the
probability of winning $10,000, assuming the gambler
initially has $1000.

6) For the insurance problem without assurance described
    in Section 4.4, find (approximately) the value d of
    dividend proportion which maximizes the expected total
    return over a 20 year operating horizon. (Simulation
    will be required.)

7) Suppose that for an inventoried stock, there are two
    possible suppliers, which will be denoted by A and B.
    Under supplier A, the process is exactly the same as in
    the second simulation study of Section 4.5; that is, the
    charge for placing an order is $15, the lead time is
    uniformly distributed on [2T/3, T], where the review
    period T is 75 days, etc. On the other hand, supplier
    B produces items which are just as effective as A's, but
    take less space and consequently have a daily holding
    cost of only $0.003 per item (instead of $0.004). But
    under supplier B, the delivery takes longer.

Specifically, the lead time is uniformly distributed on
the interval [3T/4, 5T/4]. Using simulation, choose the
most desirable supplier and give an approximation to the
optimal ordering level R_A^* and R_B^* in each case.

8) Assume a warehouse supplies goods to five retail stores,
each of which has costs, review periods, and demand
exactly as in the second simulation study of Section
4.5. Each retailer uses R* = 310 as ordering level.
Assume that the demand at each retailer is statistically
independent of the others and simulate to approximate
the optimal ordering level for the warehouse, assuming
that its review period is 4T = 300 days and its daily
holding cost is $0.002 per item, its lead time is uni-
form on [T, 4T], but other constants are as in the simulation example. Find an approximately optimal ordering
level for the warehouse.

REFERENCES FOR CHAPTER 4

Casti, J., and R. Kalaba, (1973). Imbedding Methods in
    Applied Mathematics. Addison Wesley, Reading, Mass.

Chernoff, H., (1972). Sequential Analysis and Optimal De-
    sign. A SIAM (Society for Industrial and Applied
    Mathematics) Publication, Philadelphia.

Chow, Y., H. Robbins, and D. Siegmund, (1971). Great Ex-
    pectations: The Theory of Optimal Stopping. Houghton
    Mifflin, New York.

DeGroot, M., (1970). Optimal Statistical Decisions. McGraw
    Hill, New York.

Dubins, L., and L. Savage, (1965). How to Gamble If You
    Must. McGraw Hill, New York.

Freeman, H., (1965). Discrete-Time Systems. Wiley, New
    York.

Freiberger, W., and U. Grenander, (1971). A Short Course
    in Computational Probability and Statistics. Springer
    Verlag, New York.

Goldberg, S., (1958). Difference Equations. Wiley, New
    York.

Gordon, D., (1969). System Simulation. Prentice-Hall,
    Englewood Cliffs, N. J.

Grenander, U., (1973). "Computational Probability and
    Statistics." SIAM Review, 15, pp. 134-192.

Hadley, G., and T. Whiten, (1963). Analysis of Inventory
    Systems. Prentice-Hall, Englewood Cliffs, N. J.

Hammersley, J., and D. Handscomb, (1964). Monte Carlo Methods. Methuen, London.

Johnson, L., and D. Montgomery, (1974). Operations Research in Prodiction, Planning, Scheduling, and Inventory Control. Wiley, New York.

Jury, E., (1964). Theory and Application of the Z-Transform Method. Wiley, New York.

Maistrov, L., (1974). Probability Theory, A Historical Sketch. Academic Press, New York.

Smith, D., Ed., (1959). A Source Book in Mathematics. Dover, New York.

Teichroew, D., and J. Lubin, (1966). "Computer Simulation: Discussion and Comparison of Languages." Comm. A.C.M., Vol. 9, pp. 723-741.

CHAPTER 5

LIMITING PROCESSES FOR RANDOM WALKS AND TIME SERIES SIMULATION

> Blind fortune rules the affairs of men, dis-
> pensing with unthinking hand, her gift oft
> favoring the worst.

Phaedra
Seneca

5.0 BACKGROUND

The Bernoulli variable is the simplest non-degenerate random variable. Recall that the outcomes of a Bernoulli experiment are 1 or 0, with probability of p and $1 - p$ respectively, $0 < p < 1$. As noted in Chapters 3 and 4, this is a natural model for coin tossing (associating a 1 with "head," 0 with "tail"). By "model" one usually means a mathematical description of a physical or "real world" phenomenon. Thus the Bernoulli variable is a popular model in the sense that it is supposed to model many naturally

occurring situations (such as coin-tossing, occurrence or
non-occurrence of a given event, passage or failure of a
test (especially of an acceptance test of a newly manufactured part), the presence or absence of a misprint on a
page of text, etc.).

The Bernoulli process (that is, a sequence of independent Bernoulli variables), besides leading us to interesting gaming phenomena, has played an even more important
role in the historical development of probability by
affording the structure for the first important central
limit theorem, the Laplace theorem, which asserts that
"standardized" sums of outcomes of Bernoulli trials converge in distribution to the standard normal variate. This
discovery provided an important motivation for using the
normal law inasmuch as, in a certain sense, it affords a
good approximation for sums of independent variates; since
the time of the discovery of central limit theorems, the
normal law has been the most prominent of all distribution
families in the literature of probability and statistics.

Closely related to the Laplace central limit theorem
is the fact that the Bernoulli process provides an approximation to Brownian motion (otherwise known as the Wiener
process), which is one of the most important stochastic
processes.

Being thus motivated by the normal law, we discuss the
multivariate normal law, presenting an efficient technique
for obtaining samples of any multivariate normal vector by
means of a linear transformation of a vector of independent
standard normal samples.

Gaussian time series are becoming an increasingly popular model in engineering and the physical sciences. Such

a time series may be viewed as an infinite dimensional normal random vector, but it is not feasible to use this insight directly as a means for generating time series observations. Instead, as we will see, it is possible to exploit
a connection between Gaussian time series and difference
equations in order to obtain a surprisingly easy and fast
algorithm for simulation of long time series segments.

5.1 A BERNOULLI PROCESS ILLUSTRATION OF LAPLACE'S THEOREM

     AND BROWNIAN MOTION

The normal (or Gaussian) random variable with parameter
(m, σ^2) has mean m and standard deviation σ, $\sigma > 0$, and is
determined by the probability density function

$$f(x) = (2\pi\sigma^2)^{-1/2}\exp(-(x - m)^2/2\sigma^2), \quad -\infty < x < \infty.$$

The normal variate is the most widely used random variable
in probability and statistics. This popularity stems from
its convenient analytic properties and from the fact, expressed precisely in the many forms of the central limit
theorem, that under fairly general conditions, "standardized"
sums of random variables converge in distribution* to a
normal variate. The reader is urged to consult (Breiman,
1968) or the more detailed (Gnedenko and Kolmogorov, 1954)

---

*That is to say, the cumulative distribution functions of
the standardized sums converge pointwise to the standard
normal cumulative distribution function.

for discussion of the central limit problem. Another source
of interest in the normal variate is due to the fact that
many workers in psychology, social science, and related
areas believe that the normal distribution accurately represents the distribution amongst the population of physio-
logical and psychological factors such as scholastic apti-
tude, as numerically measured by certain tests.

Our attention in this chapter will first turn to experimental study of the Laplace Theorem*. This is a special
case of the central limit theorem which, in the words of
L. Breiman (1968, p. 9), "along with the law of large numbers shares the throne of probability theory." M. Loève
(1955, p. 168) writes, "The oldest and, until recent years,
almost the only general problem of probability theory is
the 'Central Limit Problem,'...." Let us formally state
Laplace's theorem.

THEOREM 5.1. Let $\{X_i\}$ be a sequence of independent Bernoulli
variables with parameter p, and define the "standardized
sum" Z_n by the formula

$$Z_n = \sum_{i=1}^{n} (X_i - p)/[np(1 - p)]^{1/2}. \qquad (5.1)$$

*Also called the Laplace-DeMoivre theorem in recognition
that DeMoivre proved it for the special case of Bernoulli
parameter 0.5.

Then $Z_n \to Y$, in distribution, Y being the normal variate
with mean zero and standard deviation 1.

Note that since X_i - p has mean zero and variance
p(1 - p), the above theorem is a special case of the version
of the central limit theorem stated in Section 2.3, in
connection with the central limit algorithm for generating
approximately normal variates. Let us assert at this point
that the Laplace theorem yields a highly inefficient normal
generator. Our interest in it is only to show how simulation can be used to illustrate the limit theorems of probability theory.

Whereas the Laplace theorem assures us that for "large
enough" n Z_n is accurately approximated by the standard
normal variate, in many applications it is important to
have some idea concerning what magnitude of n is large
enough. There is some abstruce mathematical theory which
provides partial answers to this question. (See especially
the Berry-Esseen theory referred to in Loève (1955, p. 282).)
But in many instances, simulation and computer analysis
provides a faster and more useful answer. Clearly, thee
Laplace theorem, like essentially any other theorem of
probability theory, is subject to computer experimentation.
Such experimentation does not substitute for a solid proof,
but it can lead to insight that cannot be gleaned through
highly analytical proofs such as abound in the literature
of limiting distributions of probability theory.

Let G_n denote the distribution function of the variable
Z_n, n = 1,2,..., defined by (5.1) above and let Φ denote
the distribution function of the normal random variable
with m = 0, σ = 1. The Laplace theorem implies (using some
standard results of analysis) that

$$\| G_n - \Phi \| \rightarrow 0, \qquad\qquad (5.2)$$

where $\| h \|$ denotes $\max_x |h(x)|$. Let $F(M, n)$ denote the empirical distribution function determined by M independent observations of Z_n. From the theory of empirical distribution functions outlined in Section 1.2, we have that with probability 1,

$$\| F(M, n) - G_n \| \rightarrow_M 0, \qquad\qquad (5.3)$$

(5.2) and (5.3) jointly imply that with probability of 1, $\| F(M, n) - \Phi \| \rightarrow 0$ as M and n increase without bound. Below we have tabulated some simulated values of $\| F(M, n) - \Phi \|$ for various values of M and n, the Bernoulli parameter p for the standardized sum Z_n being held at 0.35. We can interpret $\| F(M, n) - \Phi \|$ as a K-S statistic for testing the normality of Z_n. Such experimentation can lead to an intuitive idea about how fast such convergence occurs.

Table 5.1

Table of K-S Statistics For Laplace's Theorem

		n			
		10	20	50	100
	10	.344	.264	.112	.155
M	50	.185	.221	.104	.079
	100	.155	.130	.124	.090

n = Number of Bernoulli Trials in the Sum

M = Number of Observations of Z_n

It turns out that most of the entries in the above
table pass the K-S test for normality. As one should anticipate for a fixed value of M, the "fit" is better for larger
n. For M = 10, the acceptance threshold is 0.32 at the 20%
level and 0.36 at the 10% level. For M = 100, the respective thresholds are 0.107 and 0.122.

We will now give a construct by which a Bernoulli time
series can be made to approximate Brownian motion. Fix
$\Delta > 0$, and construct the interpolation of random sums

$$S_t(\Delta) = h(\Delta)\Sigma\, X_i: \quad i \leq t/\Delta,\ 0 \leq t < \infty, \qquad (5.4)$$

where X_i's are Bernoulli variables scaled and translated so
that they have mean zero and magnitude 1. Under proper
choice of the scalar $h(\Delta)$, $S_t(\Delta)$ converges to Brownian motion, another of the most important stochastic processes.

Specifically, we assume that $\{X_i\}$ is a sequence of
independent random variables such that

$$P[X_i = 1] = P[X_i = -1] = 1/2.$$

Notice that the graph of $S_t(\Delta)$ is a scaled random walk of
the type discussed in Chapters 3 and 4. The scalar $h(\Delta)$
in (5.4) is selected so that regardless of the value Δ,
the variance of $S_t(\Delta) = t$. This determines $h(\Delta)$ as follows:

$$\text{Variance}(S_t(\Delta)) = t = (h(\Delta))^2 \sum_{i \leq t/\Delta} \text{variance}(X_i)$$

$$= [\text{integer part of } (t/\Delta)](h(\Delta))^2 \approx t/\Delta(h(\Delta))^2.$$

Thus,

$$h(\Delta) = \Delta^{1/2}.$$ (5.5)

Notice that by Laplace's theorem,

$$N^{-1/2}(\sum_{i=1}^{N} X_i) \to Y_1, \text{ in distribution}$$ (5.6)

where here and below, Y_r denotes the normal random variable with parameter $(0, r)$. In view of (5.6), if we define $N = t/\Delta$, then $h(\Delta) = (t/N)^{1/2}$; and letting Δ tend to 0,

$$S_t(\Delta) = h(\Delta) \sum_{i=1}^{N} X_i = (t/N)^{1/2}(\sum_{i=1}^{N} X_i) \to Y_t.$$ (5.7)

More generally, since the X_i's are independent, for $t_1 < t_2 < \cdots < t_n$,

$$S_{t_1}(\Delta), (S_{t_2}(\Delta) - S_{t_1}(\Delta)), \ldots, (S_{t_n}(\Delta) - S_{t_{n-1}}(\Delta))$$

are statistically independent and, by the reasoning used in establishing (5.7), converge respectively to

$$Y_{t_1}, Y_{t_2-t_1}, \ldots, Y_{t_n-t_{n-1}}.$$

From these considerations, we have established that the stochastic processes $S_t(\Delta)$ converge, as $\Delta \to 0$, to (normalized) <u>Brownian motion</u> S_t. Brownian motion, a

stochastic process defined on the time set of non-negative
numbers, is uniquely determined by the following properties:

1) $E[S_t] = 0$, for all t.
2) For any times $t_1 \leq t_2 \leq \cdots \leq t_n$

$$S_{t_1}, (S_{t_2} - S_{t_1}), \ldots, (S_{t_n} - S_{t_{n-1}})$$

are statistically independent and normally distributed with parameters $(0, t_1)(0, t_2 - t_1), \ldots,$
$(0, t_n - t_{n-1})$, respectively.
3) $S_0 = 0$.

The above development follows Feller, p. 354f, who uses
"diffusion" to denote Brownian motion.

In Figure 5.1, we have plotted $S_t(\Delta)$, $t \leq 1$, for path
associated with $\Delta = .001$.

As proven in Breiman (1968, Chapter 12), sample paths
of Brownian motion have some exotic properties. For example,
with probability 1, a Brownian motion function is continuous
everywhere, but does not have a derivative at any point.
Furthermore, in any neighborhood of t, there are infinitely
many $t_i \neq t$, such that $S_{t_i} = S_t$. Thus, for example, with
probability 1, $S_t = 0$ for infinitely many t in $(0, 1)$, since
$S_0 = 0$.

Brownian motion was proposed by Bachelier and Einstein
to describe motion in one dimension of a small object
buffeted at random by molecules and other small objects.
Such motion was first observed by the biologist Brown (in
1826), who watched a small particle under a microscope.

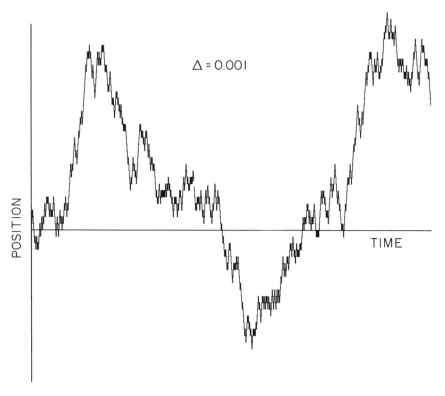

Figure 5.1   Random Walk Approximation of Brownian Motion

Of course, what he saw was Brownian motion in two dimensions.
That is to say he observed the points traced out by ordered
pairs (R_t, S_t), where R_t and S_t are independent Brownian
motions. Presumably, the path he saw resembled the plot
shown in Figure 5.2, which gives the graph of two indepen-
dent simulated paths $S_t(\Delta)$ and $S'_t(\Delta)$ with $\Delta = 10^{-4}$. The
figure linearly interpolates the points $\{(S_t(\Delta), S'_t(\Delta)):$
$t = 25n\Delta: \ 1 \leq n \leq 2000\}$.

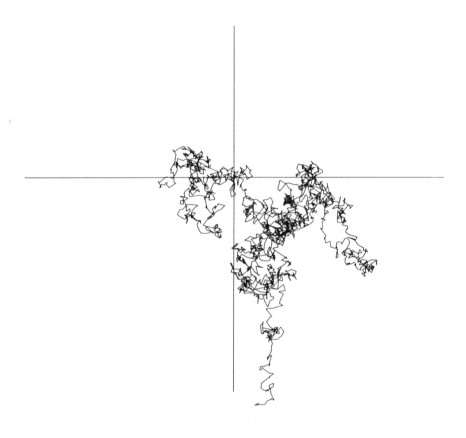

Figure 5.2   A Random Walk Approximation of Brownian Motion

in the Plane

5.2 MULTIVARIATE NORMAL GENERATION

The dimension n having been fixed, the multivariate
normal distribution is determined by the parameters (μ, Σ)
where Σ is an nth order symmetric positive definite matrix
and μ is an arbitrary n-tuple. The density function of
the multivariate normal (μ, Σ) random vector is defined for
each real n-tuple $\underline{x} = (x_1, \ldots, x_n)^T$ by

$$f(\underline{x}) = \det(\underline{\Sigma})^{-1/2}(2\pi)^{-n/2}\exp(-\frac{1}{2}(\underline{x} - \underline{\mu})^T\underline{\Sigma}^{-1}(\underline{x} - \underline{\mu})).$$

In the above expression the superscript T signifies "trans-
pose" and "det" denotes "determinant of."

There are two important properties of this distribution
which are particularly cogent to the simulation technique
to be revealed.

i) If $\underline{A}$ is an nth order non-singular matrix, $\underline{b}$ an
arbitrary n-tuple, and $\underline{X}$ is a multivariate normal random
vector with parameter $(\underline{u}, \underline{\Sigma})$, then the random vector

$$\underline{Y} = \underline{A}\underline{X} + \underline{b}$$

is multivariate normal with parameter $(\underline{b} + \underline{A}\mu, \underline{A\Sigma A}^T)$.

ii) If $\underline{\Sigma}$ is an arbitrary symmetric positive definite
matrix, the equation

$$\underline{\Sigma} = \underline{A A}^T \tag{5.8}$$

has a solution.

All these facts are derived in Section 7.4 of Wilks
(1962) as well as Section 7.1 of Lindgren.

The matrix $\underline{\underline{\Sigma}}$ turns out to be the <u>covariance matrix</u>
of the multivariate normal law. That is the (i, j) coordi-
nate of $\underline{\underline{\Sigma}}$ is $E[(X_i - \mu_i)(X_j - \mu_j)]$. Thus it is clear that
to the extent that one can generate independent univariate
normal (0, 1) samples (by the methods of Section 2.3, for
example), one can generate n-variate normal $(\underline{0}, \underline{\underline{I}})$ samples
($\underline{\underline{I}}$ being the identity matrix and $\underline{0}$ the zero vector) by
collecting in vectors the successive blocks of n univariate
normal outcomes. Let $\underline{X}$ be a multivariate normal $(\underline{0}, \underline{\underline{I}})$ random vector. Then from facts i) and ii) above, it is clear
that if $(\underline{\mu}, \underline{\underline{\Sigma}})$ is an arbitrary multivariate normal parameter
and $\underline{\underline{A}}$ satisfies the relation $\underline{\underline{\Sigma}} = \underline{\underline{A}}\underline{\underline{A}}^T$, then

$$\underline{Y} = \underline{\underline{A}}\underline{X} + \underline{\mu}$$

is multivariate normal with parameter $(\underline{\mu}, \underline{\underline{\Sigma}})$.

There are a multitude of possible ways to compute
representations satisfying (5.8). Among the undesirable
methods are those which require finding the eigenvectors of
$\underline{\underline{\Sigma}}$. The numerical technique which seems most appealing to
this author is <u>Choleski's method</u> (described in Dahlquist
and Björck (1974), p. 158) which gives a matrix $\underline{\underline{A}} = \{a_{ij}\}$
which is lower triangular through solving the system of
$n(n + 1)/2$ linear equations

$$a_{kj} = (\underline{\underline{\Sigma}}_{kj} - \sum_{p=1}^{j-1} a_{kp}a_{jp})/a_{jj}, \quad j = 1,2,\ldots,k - 1$$

$$a_{kk} = (\underline{\underline{\Sigma}}_{kk} - \sum_{p=1}^{k-1} a_{kp}^2)^{1/2}.$$

5.3 SIMULATION OF STATIONARY GAUSSIAN TIME SERIES

Time series (a synonym for sequences of random variables) are playing an increasing prominent role in modern
engineering. We let $\{X_i\}$ denote a time series. The range
of i is generally the set of all integers (positive and
negative). From advanced probability theory, it is known
that such sequences are uniquely determined by distribution
functions which give the probabilities of the random vectors
$(X_{-t}, X_{-t+1}, \ldots, X_t)$ for every $t \geq 1$. Random sequences of
such generality are hard to work with because of insurmountable problems arising in their probabilistic analysis
and statistical inference. Stationary processes constitute
an important subclass of time series because they are analytically and computationally tractible. A time series is
said to be stationary if for each pair of integers N and M,
$M > 0$, the distribution of $(X_1, \ldots, X_M)$ coincides with the
distribution of $(X_{N+1}, X_{N+2}, \ldots, X_{N+M})$. A random sequence is a
Gaussian time series if each finite collection of variables
$X_{n_1}, X_{n_2}, \ldots, X_{n_t}$ is multivariate normal. Stationary Gaussian
time series have been used to model dynamic phenomena such
as communication signals, prices of commodities, annual pre-
cipitation data, chemical concentrations, and daily river
flows. In this section, we describe simulation methods for
stationary Gaussian time series. We remark that since their
finite dimensional projections are normal, stationary
Gaussian time series are completely determined by the common

mean $\mu = E[X_1]$ and the function $R(n) = E[X_n X_0] - \mu^2$, which

is known as the <u>covariance function</u> of the process. We

commend to the reader Parzen (1962) and Box and Jenkins

(1970) for supplementary discussion of the nature and use

of time series. Rozanov (1967) and Parzen (1967) are

scholarly compendiums on this subject.

   Our intention in this section is to present a pre-

scription and rationale for simulation of finite segments

of stationary Gaussian time series. The most obvious simu-

lation procedure is to employ the multivariate simulation

technique of Section 2.4 to successively simulate X_1 to be

normal with parameter $(\mu, R(0))$, X_2 to be the normal condi-

tioned variable determined by the parameter

$$\left[\binom{u}{u}, \begin{pmatrix} R(0) & R(1) \\ R(-1) & R(0) \end{pmatrix}\right]$$ and conditioned by the X_1 observation,

and in general, X_n the conditioned variable determined by

the covariance function values $R(m)$, $|m| \leq n$, and condi-

tioned by the already observed vector $(X_1,\ldots,X_{n-1})$. This

scheme turns out to be computationally unsatisfactory for

even moderately large n because the distribution of the con-

ditioned variable $X_n|X_{n-1},\ldots,X_1$ involves algebraic opera-

tions on the <u>nth</u> order covariance matrix $\{\underline{\Sigma}_{ij}\}_{1\leq i,j\leq n}$

$= \{R(i - j)\}$, which has n^2 coordinates. We discuss a much

more fruitful approach below.

   For simplicity of notation, we assume in the following

that the mean of the time series to be discussed are zero.

The reader will see the easy modification needed for the

general case.

   In order to provide a setting for a satisfactory simu-

lation algorithm, it is necessary to delve slightly more

deeply into the properties of stationary time series. Let

R_X denote the covariance function for some time series $\{X_n\}$

and let $\{c(i)\}$ be some scalar sequence. We determine the time series $\{Y_m\}$ according to the convolution operation

$$Y_m = \sum_i c(i) X_{m-i}. \tag{5.9}$$

Then R_Y, the covariance function of the Y time series, can be expressed in terms of the covariance R_X and the $c(i)$'s as follows:

$$R_Y(m) = E[Y_m Y_0] = E[\sum_{i,j} c(i) X_{m-i} c(j) X_{-j}]$$

$$= \sum_{i,j} c(i) c(j) R_X(m - i + j). \tag{5.10}$$

The idea in the method to follow is, given a desired covariance R, to generate a random process $\{X_i\}$ which is computationally "easy" and then find a sequence $c = \{c(i)\}$ such that the Y process obtained through (5.9) has a co-variance R_Y which equals or approximates R.

The reader will recognize the last term in (5.10) as being a double convolution. In this case, as in many other instances, it is advantageous to use the fact that convolu-tion in the time domain corresponds to multiplication in the Fourier transform domain. We will use only the most elemen-tary properties of Fourier series, the theory of which is described in (Apostol, 1957), for instance. Let $A = \{A(i)\}$ be any sequence such that $\sum |A(i)| < \infty$. We define its Fourier transform by the formula

$$F(A) = f_A(\omega) = (1/2\pi) \sum_n A(n) \exp(-i\omega n) \tag{5.11}$$

where ω ranges over the interval $[-\pi, \pi]$. The convergence
of the series (5.11) for each ω is assured by the absolute
summability assumption above. We will assume that every
covariance function R under discussion is similarly summable;
that is, $\Sigma\, R(n) < \infty$.

Let A, B, R, and c denote sequences and let F denote
the Fourier transform operation defined in (5.11). (Actually, F is the inverse transform for Fourier series.) Then,
as the reader may verify, if $*$ denotes "convolution" and $\overline{f}$
denotes the complex conjugate of f, letting f_A denote F(A),
and $f_B = F(B)$, we have

$$F(A * B) = f_A f_B,$$

and

$$F(\{A(-n)\}) = \overline{f}_A.$$

From these rules, we have immediately from (5.10) that since
$R_Y = c * R_X * \{c(-n)\}$, if f_X, f_Y, and f_c denote the trans-
forms of R_X, R_Y, and $\{c(n)\}$ respectively,

$$f_Y = f_c \overline{f}_c f_X. \qquad\qquad (5.12)$$

In the discussion to follow, R will denote the covariance function of the desired Gaussian time series and f its
Fourier Transform F(R). Such a transform f is customarily
referred to as a <u>spectral density</u>. We will now make the

assumption that f is a <u>rational spectral density</u>*, an assumption commonly found in the applications literature. This
assumption is supported by the obvious fact that if g is
the transform of any summable covariance function whatsoever,
and ε is a positive number, there is a rational spectral
density $r(\omega)$ such that

$$\max_{\omega} |g(\omega) - r(\omega)| < \varepsilon. \tag{5.13}$$

In view of our summability assumption, an obvious
demonstration of (5.13) is to let $f_m(\omega)$ be the transform of
R_m where $R_m(n) = R(n)$, $|n| \leq m$ and $R_m(n) = 0$, otherwise.
Clearly f_m is a rational function of $e^{i\omega}$. Then from
Parseval's equation (Apostol, 1957, p. 466),

$$\frac{1}{2\pi} \int_{-\pi}^{\pi} |f_m - g|^2 \, d\omega = \sum_{|n|>m} R(n)^2 \to_m 0.$$

As is demonstrated in Rozanov (1967) as well as in
other books on stationary stochastic processes, if f is a
rational spectral density, then f admits the polynomial
factorization

$$f(\omega) = (P_B(e^{i\omega})\overline{P_B}(e^{i\omega}))/P_A(e^{i\omega})\overline{P_A}(e^{i\omega})), \tag{5.14}$$

---

*That is, $f(\omega) = P_1(e^{i\omega})/P_2(e^{i\omega})$, where $P_1(z)$ and $P_2(z)$
are polynomials defined on the complex variable z.

where the roots z_i of the polynomials $P_A(z)$ and $P(z)$ lie outside the unit circle. As before, the bars in (5.14) denote "complex conjugate."

A final fact we shall need is that if $\{X_i\}$ is a time series composed of independent standard normal variates, then $R_X(0) = 1$ and $R_X(n) = 0$, $|n| > 0$. Consequently, $f_X(\omega) = 1$, for all ω. We will obtain our sequence $\{Y_i\}$ with the desired spectral density f by performing determistic operations on this independent standard normal sequence $\{X_i\}$.

With reference to (5.14), let $\{a_k\}$ $k = 0,1,\ldots,N$ and $\{b_j\}$ $j = 0,1,\ldots,M$ denote the coefficients of the polynomials P_A and P_B respectively. If we determine a time series $\{Z_m\}$ by the rule

$$Z_m = b_0 X_m + b_1 X_{m-1} + \cdots + b_M X_{m-M}, \qquad (5.15)$$

then from (5.10) and (5.12) we see that

$$f_Z = P_B(e^{i\omega}) f_X \overline{P_B(e^{i\omega})} = P_B(e^{i\omega}) \overline{P_B(e^{i\omega})}. \qquad (5.16)$$

Now we will define our output sequence $\{Y_n\}$ by the rule

$$Y_n = a_0^{-1}(-a_1 Y_{n-1} - a_2 Y_{n-2} - \cdots - a_N Y_{n-N} + Z_n). \qquad (5.17)$$

From (5.9) and (5.12) we conclude that

$$P_A * Y = Z$$

or

$$P_A * R_Y * \{P_A(-n)\} = R_Z$$

and thus

$$P_A(e^{i\omega}) f_Y \overline{P_A}(e^{i\omega}) = f_Z(\omega) = P_B(e^{i\omega}) \overline{P_B}(e^{i\omega})$$

from whence, as desired,

$$f_Y(\omega) = (P_B(e^{i\omega}) \overline{P_B}(e^{i\omega})) / (P_A(e^{i\omega}) \overline{P_A}(e^{i\omega})) = f(\omega).$$

The reason for choosing P_A so that its roots lie outside the unit circle is that this assures that (5.17) represents a stable difference equation. We summarize these developments in a formal algorithm.

ALGORITHM 5.1   A GAUSSIAN TIME SERIES SIMULATOR.

Input.   A covariance function $R(n)$ whose spectral density f is the rational function of $e^{i\omega}$ whose components are denoted as in (5.14), with

$$P_A(z) = \sum_{i=0}^{N} a_i z^i, \quad P_B(z) = \sum_{i=0}^{M} b_i z^i.$$

1)   Simulate a sequence $\{X_i\}_{i>0}$ of independent standard normal variates (according to methods in Section

2.3, for example).

2) Determine a sequence $\{Z_n\}$ $n \geq M$ by the rule

$$Z_n = \sum_{i=0}^{M} b_i X_{n-i}.$$

3) Let N' denote the maximum of N and M, and determine $\{Y_n\}$ by the nonhomogeneous linear difference equation

$$Y_n = a_0^{-1}(-\sum_{i=1}^{N} a_i Y_{n-i} + Z_n), \quad n \geq N'$$

(The initial values $Y_0, Y_1, \ldots, Y_{N'}$ may be selected arbitrarily.*)

Output. $\{Y_n\}$, a Gaussian time series having the desired covariance function R.

In time series parlance, the process $\{Z_n\}$ is called a moving average of the $\{X_i\}$ process. The $\{Y_i\}$ process is known as an autoregressive process because Y_n is determined by a regression equation on the earlier Y values. With reference to the $\{X_i\}$ process, the $\{Y_n\}$ process is called an autoregressive moving average process (ARMA) and is a popular model in the applications literature. From our

---

*Strictly speaking $(Y_0, Y_1, \ldots, Y_{N'})$ should be chosen to be multivariate normal with $\underline{0}$ mean and covariance matrix determined by $\underline{\Sigma}_{ij} = R(i - j)$, but the lack of exactitude here will not make a significant difference for moderately long chains.

analysis, we have seen that any time series having a spectral density which is a rational function of $e^{i\omega}$ is an ARMA
process, and further, in view of (5.13), that any stationary
Gaussian time series can be approximated arbitrarily accurately by an ARMA process.

We have discussed the simulation of Gaussian time
series. Newman and Odell (1971) give a parallel approach
to simulating continuous-time Gaussian stationary processes
at any discrete set of time points.

A computer simulation was performed to illustrate step
2 of the preceding algorithm. In our study, we set $b_0 = 1$,
$b_1 = 1$, $b_2 = 2$, and $b_3 = 3$. M is taken to be 3. The time
series $\{X_i\}$ in step 1 of the algorithm was taken to be an
independent sequence of standard normal variates obtained
by the Box-Muller algorithm. By evaluating the products
of the appropriate moving averages, one may readily evaluate
$R_Z(j)$. In our example, for $0 \le j \le 3$, $R_Z(j) = \sum_{v=0}^{3-j} b_v b_{v+j}$,
and these values are presented in Table 5.2 along with the
sample covariance function $\overline{R}_Z(j)$ defined for each integer
j less than the simulated sample length n by the formula

$$R_Z^n(j) = (1/(n - j)) \sum_{v=1}^{n-j} Z_v Z_{v+j}. \qquad (5.18)$$

In (5.18) the Z_i's are values in a simulated time series of
length n. According to ergodic theorems in time series
analysis (Rozanov (1967), Section 1.6) for each j, $R_Z^n(j)$
$\rightarrow R_Z(j)$ as the sample length n increases without bound. In
Table 5.2, we note that for this simulation in which the
sample length was 200, the agreement between the sample and
the actual covariance functions is not bad.

Table 5.2

Comparison of Actual and Sample Covariance Functions For a

Moving Average

j	Sample Covariance Function $R_Z^n(j)$	Actual Covariance Function $R_Z(j)$
0	17.4	15.0
1	10.2	9.0
2	7.1	5.0
3	6.1	3.0

Coefficients: $b_0 = 1$, $b_1 = 1$, $b_2 = 2$, $b_3 = 3$

Length n of Simulated Series: 200

A second time series simulation was performed in order to illustrate Step 3 of the time series simulation algorithm. For ease in finding the theoretical covariance function, we took $Z_n = X_n$, the standard normal sequence of Step 1. The parameters of Step 3 are N = 1, $a_0 = 1$, and $a_1 = 0.75$. Then one may confirm that for each integer j, the theoretical covariance function of the time series is given by

$$R_Y(j) = (a_1)^{|j|}/(1 - a_1^2).$$

In Table 5.3, we have tabulated these theoretical values along with the sample covariance functions of the simulated sequence $\{Y_i\}$, the sample function being based on a

simulated time series of length 1,000. The time series
determined by a linear difference equation as in Step 3 is
called an <u>autoregressive process</u>.

Table 5.3

Comparison of Actual and Sample Covariance Functions for an

Autoregressive Process

j	Sample Covariance Function $R_Z{}^n(j)$	Actual Covariance Function $R_Y(j)$
0	2.36	2.29
1	1.82	1.71
2	1.40	1.29
3	1.07	.964
4	.819	.723

EXERCISES

1) (For readers with background in hypothesis testing.)
    With respect to the test statistic $\| F(M, n) - \Phi \|$ de-
    fined in connection with Table 5.1, justify the following
    assertions:

    a) If n and the test level α are held fixed but M be-
        comes large, the probability of rejection converges
        to 1.

    b) On the other hand, if M is held fixed and n in-
        creases without bound, the probability of rejection
        will approach α.

2) Let $\{X_i\}$ be an independent, identically distributed
    sequence of random variables with common finite variance
    σ^2 and mean μ. Define the standardized sum Z_n
    $= \sum_{i=1}^{n} (X_i - \mu)/\sqrt{n}\ \sigma$. According to the central limit
    theorem, as n increases, Z_n converges in distribution
    to the standard normal variable. Take X to be uniform
    and use simulated observations of Z_j to form the test
    statistic $\| F(M, n) - \Phi \|$. With these values construct
    a counterpart to Table 5.1. Encircle those entries

which are rejected by the K-S test at the 10% level.
Note: Such experimentation can be useful in determining
if the normal variate is an adequate approximation for
sums of variables. This experiment is cogent to the
central limit normal variate generator given in Section
2.3.

3) Brownian motion is thought to provide a model for diffusion. An intuitive property of diffusion in a bounded
region is that as time goes on, particles tend to become
uniformly spaced about the region. We can approximate
diffusion constrained to the bounded interval [-1, 1],
for example, by letting -1 and 1 be reflecting barriers.
Thus if $S_t(\Delta) = 1$, for instance, then we will require
that the next X value be -1, and similarly, if $S_t(\Delta) = 1$,
the next X value must be +1. Otherwise the $S_t(\Delta)$
approximation of Brownian motion is as described in
equation (5.7). Set $\Delta = 10^{-3}$, and repeatedly simulate
the value $S_2(\Delta)$. Are these observations approximately
uniformly distributed on [-1, 1]? Graph them.

4) Define $\underline{0} = (0, 0, 0)^T$, I the third-order identity matrix
and

$$A = \begin{bmatrix} 1 & 4 & 3 \\ 2 & 1 & 1 \\ 0 & 1 & 4 \end{bmatrix},$$

and generate multivariate normal observations $Y_1, Y_2, \ldots,$
Y_n of the variable Y = AX, where X is multivariate
$(\underline{0}, I)$. Letting $Y_j(k)$ denote the kth coordinate of Y_j,
compute the sample covariance

$$\text{cov}(i, k)_n = 1/n \sum_{v=1}^{n} Y_v(i)Y_v(k), \quad 1 \leq i, k \leq 3.$$

Check to see that as n increases, $\text{cov}(i, k)_n$ converges to its theoretical value, namely the (i, k) coordinate of AA^T, for each pair of values (i, k).

5) Let $\{X_j\}$ be a sequence of independent normal (0, 1) variates and

$$Z_n = \sum_{j=0}^{M} b_j X_{n-j}.$$

Find the covariance function $R_Z(j)$ in terms of the b_j's for this moving average time series and verify the third column in Table 5.2.

6) Let $\{X_j\}$ be an independent sequence of standard normal variates as in problem 5 and define

$$Y_n = -aY_{n-1} + X_n.$$

Calculate the covariance function $R_Y(j)$ as a function of a and verify the third column of Table 5.3.

7) Let $\{X_n\}$ be an independent sequence of standard normal variables and define $\{Y_n\}$ to be the autoregressive sequence determined by

$$Y_n = a_0^{-1}(\sum_{i=1}^{N} a_i Y_{n-i} + X_n).$$

where the a_i's are such that the above is a stable difference equation (the roots of the associated polynomial $P_A(z)$ lie outside the unit circle). One may use the Wold representation (Rozanov, 1967, p. 56) to rigorously confirm that the Y_n time series bears the representation

$$Y_n = \sum_{i=-\infty}^{n} c_{n-i} X_i$$

for some sequence of constants $\{c_n\}$. Use this fact to establish that

a) $E[X_n Y_{n-k}] = 0$, for $k > 0$ and demonstrate that therefore the covariance function R_Y of $\{Y_i\}$ satisfies

b) $\sum_{i=0}^{N} a_i R_Y(k - i) = 0$, for $k > 0$.

8) (Continuation) Suppose a time series $\{Y_n\}$ is known to satisfy

$$Y_n = -\sum_{i=1}^{N} a_i Y_{n-i} + X_i$$

where the $\{X_i\}$ is an independent sequence of standard normal variables, and suppose further the a_i values are not known. The relationship given in part b of Problem 7 allows us to <u>identify</u> the coefficients $a_1,\ldots,a_n$ of an observed autoregressive time series from the sample covariance function (defined in connection with Tables 5.2 and 5.3). Specifically, suppose that $\overline{R}_Y(n)$ is an accurate approximation of the covariance function for the observed time series $\{Y_i\}$ having known autoregressive order N. Then by letting $k = 1,2,\ldots,N$, one obtains linear equations

$$\overline{R}_Y(1) + a_1\overline{R}_Y(0) + \cdots + a_N\overline{R}_Y(-N) = 0$$

$$\overline{R}_Y(2) + a_1\overline{R}_Y(1) + \cdots + a_N\overline{R}_Y(-N + 1) = 0$$

$$\vdots \qquad\qquad \vdots \qquad\qquad \vdots$$

$$\overline{R}_Y(N) + a_1\overline{R}_Y(N - 1) + \cdots + a_N\overline{R}_Y(-1) = 0$$

which may be solved for $a_1, \ldots, a_N$. These equations are known (Box and Jenkins, 1970, p. 55) as the Yule-Walker equations. Take $N = 2$, $a_0 = 1$, $a_1 = -5/6$, $a_2 = 1/6$, and simulate a chain of length 5000. Using the Yule-Walker equations, see how closely you can recover these parameters by using the sample covariance function as the approximation $\overline{R}(j)$.

REFERENCES FOR CHAPTER 5

Apostol, T., (1957). Mathematical Analysis. Addison-Wesley, Reading, Massachusetts.

Box, G., and G. Jenkins, (1970). Time Series Analysis Forecasting and Control. Holden Day, San Francisco.

Breiman, L., (1968). Probability. Addison-Wesley, Reading, Massachusetts.

Dahlquist, G., and A. Björck, (1974). Numerical Methods. Prentice Hall, Englewood Cliffs, New Jersey.

Gnedenko, B., and A. Kolmogorov, (1954). Limit Distributions for Sums of Independent Random Variables. Addison-Wesley, Reading, Massachusetts.

Loève, M., (1955). Probability Theory. Van Nostrand, Princeton, New Jersey.

Newman, T., and P. Odell, (1971). The Generation of Random Variates. Griffin, London.

Parzen, E., (1962). Stochastic Processes. Holden Day, San Francisco.

Parzen, E., (1967). Time Series Analysis Papers. Holden Day, San Francisco.

Rozanov, Y., (1967). Stationary Random Processes. Holden Day, San Francisco.

Wilks, S., (1962). Mathematical Statistics. Wiley, New York.

CHAPTER 6

MONTE CARLO INTEGRATION AND SOLUTION OF DIFFERENTIAL EQUATIONS

> But who can teach the thoughts we should
> roll-call
> When morning finds us marching to the wall
> Under the stage direction of some goon
> Political, some uniformed baboon?
> We'll think of matters known to us,
> Empires of rhyme, Indies of calculus.
>
> <u>Pale Fire</u>
> Nabokov

6.0 BACKGROUND

The ostensible objective of this chapter is to show the
reader how simulation techniques we have been studying can
be used to evaluate integrals in a manner that is distinctly
different from the procedures of numerical analysis texts.
The author feels that the major benefits of this study lie
not only in gaining a new tool for the prosaic task of evaluating an integral, but in seeing how probability theory is

related to the integral calculus. This discussion should
help the reader to understand that in a strict sense, the
theory of integration and probability theory coincide. (A
brief but fruitful discussion of Riemann and Riemann-
Stieltjes integration is to be found in Lindgren, p. 101ff.)
So that we may get at the heart of the matter, this chapter
concludes with a brief appendix giving a specialized discussion for readers who have studied measure theory.

Monte Carlo integration has practical value as a numerical analysis procedure as well as the didactic merit we
have alluded to. In Section 6.5, highly efficient but less
intuitive Monte Carlo schemes are reported, and error bounds
are derived. From these theoretical explorations, the
reader will be able to understand why, in some circumstances,
Monte Carlo integration is competitive and even superior to
classical quadrature techniques.

The Monte Carlo method is particularly attractive for
multivariate integration because, for a fixed accuracy, the
computational effort is less sensitive to the dimension of
the domain of integration than classical quadrature schemes.
These matters are discussed in Section 6.6, where the principles of the "quasi-Monte Carlo" method--a hybrid of Monte
Carlo and quadrature techniques--is sketched. In the con-
cluding section, another link between the Monte Carlo method
and analysis is described, namely a simulation approach for
solving partial differential equations. We will present in
detail (and with a sample program) a Monte Carlo method for
solving an important boundary-value problem (the Dirichlet
problem) arising in the theory of elliptic differential
equations.

6.1   HIT OR MISS MONTE CARLO INTEGRATION

Suppose that $g(x)$ is a piecewise continuous real function such that $0 \leq g(x) \leq 1$ if x is in the unit interval
and suppose further that we wish to evaluate $\int_0^1 g(x)\ dx$.
Consider Figure 6.1 below.

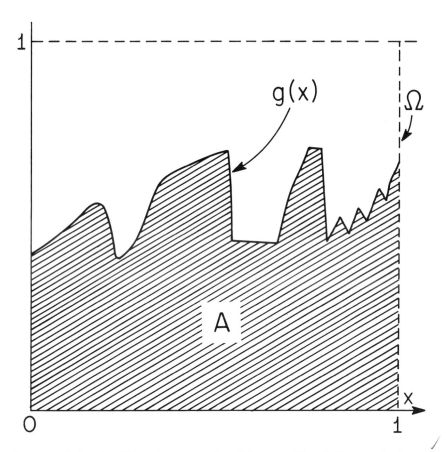

Figure 6.1   An Illustration for "Hit or Miss" Monte Carlo

Ω denotes the unit square and A = $\{(x, y): y \leq g(x)\}$. Observe that $\int_0^1 g(x) dx$ = Area under g(x) = Area(A). If W and V are independent random variables uniformly distributed on the unit interval, then for any event A

$$P_{W,V}[A] = \iint_A 1 \, dx \, dy = \text{Area}(A). \tag{6.1}$$

Imagine that we take n independent observations of the bivariate random variable (W, V), and we let n_A denote the number of these observations which are members of event A. n_A/n is the <u>hit or miss</u> estimator of $\int_0^1 g(x) dx$. The law of large numbers (Theorem 3.1) implies that with probability 1, the relative frequency of occurrence of an event converges to the probability of that event. That is, with probability 1,

$$\lim_{n \to \infty}(n_A/n - P[A]) = 0. \tag{6.2}$$

If, in the unit interval, the integrand f(x) has maximum value M greater than 1, we apply the hit or miss algorithm to the function f(x)/M, and then multiply the estimate by M. The preceding analysis implies that the following algorithm gives an estimate which converges to the desired integral. We defer to Section 6.4 an analysis of the error $|n_A/n - \int_0^1 g(x) dx|$ associated with this algorithm.

ALGORITHM 6.1 HIT OR MISS MONTE CARLO INTEGRATION ALGORITHM.

Input. A sequence $\{u_j\}_{j=1}^{2n}$ of 2n random numbers.
   1) Arrange the random numbers into n pairs,
       $\{(w_i, v_i)\}_{i=1}^{n}$ in any fashion such that each random
       number u_i is used exactly once.
   2) Say that the pair (w_i, v_i) is in the event A if
       $v_i \leq g(w_i)$. Let n_A denote the number of pairs
       (w_i, v_i) in A.

Output. n_A/n, our estimate for $\int_0^1 g(x)\,dx$.

In a computer study, we used Algorithm 6.1 to evaluate
the integrals over the unit interval of the following four
functions:

$$f_1(x) = 4(1 - x^2)^{1/2} \qquad\qquad f_3(x) = x^3$$

$$(6.3)$$

$$f_2(x) = \sin(x) \qquad\qquad f_4(x) = e^x.$$

The results of this computation are given in Table 6.1.

6.2 MONTE CARLO INTEGRATION THROUGH SAMPLE MEAN CONVERGENCE

In the method proposed here, we no longer assume that
g(x) is a bounded function; as before, we wish to find
$\int_0^1 g(x)\,dx$.
   Let X be a random variable with probability density
function $f_X(x)$. Then, by the definition of expectation,

Table 6.1

Hit or Miss Monte Carlo Integration

$$\int_0^1 4\sqrt{1 - x^2}\ dx = 3.14159$$

N	Approximation	Error
20	2.8000	.342
50	3.0400	.102
500	3.1600	.184E-1
10000	3.1172	.244E-1

$$\int_0^1 \sin(x)\ dx = .45970$$

N	Approximation	Error
20	.4000	.597E-1
50	.5000	.403E-1
500	.4560	.370E-2
10000	.4555	.420E-2

$$\int_0^1 x^3\ dx = .2500$$

N	Approximation	Error
20	.1500	.100
50	.3000	.500E-1
500	.2460	.400E-2
10000	.2433	.670E-2

Table 6.1 Continued

$$\int_0^1 e^x \, dx = 1.71828$$

N	Approximation	Error
20	1.7669	.486E-1
50	1.8484	.130
500	1.7071	.112E-1
10000	1.7187	.420E-3

$$E_X[g(X)] = \int_{-\infty}^{\infty} g(x) \ f_X(x) \ dx. \qquad (6.4)$$

Thus if U is the random variable uniformly distributed on the unit interval, then recalling that the density function for U is 1 for all x in the unit interval and 0 elsewhere,

$$E_U[g(U)] = \int_{0}^{1} g(x) \ dx. \qquad (6.5)$$

From (6.5) it is clear that G = g(U) is an induced random variable having expectation $\int_0^1 g(x) \ dx$. We now draw on the law of large numbers (Theorem 3.1): If G is any random variable having a finite mean, E[G], and if $\bar{g}_n$ is computed from independent observations of G by adding the observations together and dividing by n (i.e. $\bar{g}_n$ is the sample mean of n observations of G), then, with probability 1.

$$\lim_{n\to\infty} \bar{g}_n = E[G]. \qquad (6.6)$$

$\bar{g}_n$ is called the sample-mean estimator of $\int_0^1 g(x) \ dx$. We reiterate the sample-mean Monte Carlo method in the following algorithm:

ALGORITHM 6.2   SAMPLE-MEAN MONTE CARLO ALGORITHM.

Input. n random numbers $\{u_i\}_{i=1}^{n}$.
   1)   Compute $g(u_i)$ $1 \le i \le n$.
   2)   Compute the sample mean, $\bar{g}_n = \dfrac{1}{n} \sum_{i=1}^{n} g(u_i)$.

<u>Output</u>. $\bar{g}_n$, our estimate for the integral $\int_0^1 g(x)\ dx$.

An error analysis of the sample-mean estimator is deferred until Section 6.4, wherein the performance of Algorithm 6.2 is compared with Algorithm 6.1. Through translation of the domain of integration to the unit interval, the above algorithm as well as the preceding may, of course, be modified to perform integration over any finite interval. We leave the details of this to the reader.

In a computer study, we applied Algorithm 6.2 to the functions f_1, f_2, f_3, and f_4 defined in connection with Table 6.1 as well as to the unbounded function $f_5 = x^{-1/2}$. The results of the calculation are given in Table 6.2. In general the error entries in this table are noticeably smaller than the corresponding errors of the hit or miss simulations recorded in Table 6.1. We will see later that there is theoretical reason for this observation.

6.3   MONTE CARLO INTEGRATION FOR IMPROPER INTEGRALS

Suppose we wish to integrate a function $g(x)$ over an aribitrary interval E. Suppose further that $f(x)$ is a probability density function such that

$$\int_E f(x)\ dx = 1.$$

Then

$$\int_E g(x)\ dx = \int_E g(x)/f(x)\ f(x)\ dx = E_X[g(X)/f(X)]. \quad (6.7)$$

Table 6.2

Sample Mean Monte Carlo Integration

$$\int_0^1 4\sqrt{1 - x^2}\ dx = 3.14159$$

N	Approximation	Error
20	3.2544	.113
50	3.1599	.183E-1
500	3.1566	.150E-1
10000	3.1346	.695E-2

$$\int_0^1 \sin(x)\ dx = .45970$$

N	Approximation	Error
20	.4048	.548E-1
50	.4398	.199E-1
500	.4512	.851E-2
10000	.4606	.912E-3

$$\int_0^1 x^3\ dx = .25000$$

N	Approximation	Error
20	.2166	.334E-1
50	.2437	.635E-2
500	.2443	.574E-2
10000	.2518	.178E-2

Table 6.2 Continued

$$\int_0^1 e^x \, dx = 1.71828$$

N	Approximation	Error
20	1.6259	.924E-1
50	1.6883	.299E-1
500	1.7036	.147E-1
10000	1.7207	.239E-2

$$\int_0^1 1/\sqrt{x} \, dx = 2.0000$$

N	Approximation	Error
20	2.0667	.667E-1
50	1.8986	.101
500	1.8921	.108
10000	1.9566	.433E-1

This observation leads to the following sample-mean algorithm for evaluating $\int_E g(x)\,dx$.

ALGORITHM 6.3   MONTE CARLO ALGORITHM FOR INFINITE INTERVALS

Input. $\{x_i\}_{i=1}^n$, n independent observations distributed
according to the common probability density function $f(x)$.

    1) Compute $g(x_i)/f(x_i)$, $1 \le i \le n$.

    2) Find the sample average $\bar{g}_n = 1/n \sum_{i=1}^n g(x_i)/f(x_i)$.

Output. $\bar{g}_n$, our estimate for $\int_E g(x)\,dx$.

In Table 6.3, we present the results of a computer run
using Algorithm 6.3 for evaluating the improper integral
$\int_0^\infty (x + 1)^{-2}\,dx$, whose value is known to be 1.   We chose the
density f to be that of the exponential law with parameter
0.1.   Exponential generation was discussed in Chapter 2.

Table 6.3

Approximation of an Improper Integral

N	Approximation	Error
20	1.131	.131
50	1.144	.144
500	1.090	.090
10000	.977	.023

If g(x) can be factored as below,

$$g(x) = k(x)f(x), \qquad (6.8)$$

where f(x) is a probability density function integrating to
1 on E, then for X a random variable having density f(x)

$$E_X[k(X)] = \int_E k(x)f(x) \ dx = \int_E g(x) \ dx. \qquad (6.9)$$

This leads immediately to the following minor modification
of Algorithm 6.3, namely: generate a sequence of observations $\{X_i\}$ having probability density function f(x). Then
(again by virtue of the law of large numbers) as $n \to \infty$

$$\bar{g}_n = \frac{1}{n} \sum_{i=1}^{n} k(X_i) \to E_X[k(X)] = \int_E g(x) \ dx. \qquad (6.10)$$

In Table 6.4, we present simulation results of using
(6.10) to evaluate $\int_0^\infty x^2 e^{-x} \ dx = \Gamma(3) = 2$. k(x) was taken
to be x^2 and f(x) is the exponential density with parameter
1.

Table 6.4

Monte Carlo Integration of a Product Function

N	Approximation	Error
20	2.219	.219
50	2.316	.316
500	1.960	.040
10000	2.041	.041

6.4   ERROR ANALYSIS FOR MONTE CARLO INTEGRATION

Let $I \equiv \int_0^1 g(x)\ dx$. In the case of hit or miss Monte Carlo, n_A (in equation (6.2)) is the sum of n independent Bernoulli trials each having parameter I. Consequently the expectation of n_A/n is I (one says that the "hit or miss" method is unbiased) and the variance of n_A (abbreviated var(n_A)) is $nI(1 - I)$. Therefore the variance of the hit or miss estimator n_A/n is given by

$$var(n_A/n) = 1/n^2 var(n_A) = I(1 - I)/n.$$

In summary, under hit or miss Monte Carlo,

$$E[(\text{hit or miss error})^2] = var(n_A/n)$$

$$= I(1 - I)/n. \qquad (6.11)$$

By elementary calculus methods, one may confirm that

$I(1 - I) \leq 1/4$ and hence always,

$$E[(\text{hit or miss error})^2] \leq 1/4n. \qquad (6.12)$$

As in the case of hit or miss Monte Carlo, sample mean Monte Carlo is unbiased; that is,

$$E[\bar{g}_n] = E[g(U)] = \int_0^1 g(x) \, dx = I.$$

Then from inspection of the definition of the sample mean estimator $\bar{g}_n$

$$E[(\text{sample mean error})^2] = E[(\bar{g}_n - I)^2] = \text{var}(\bar{g}_n)$$

$$= 1/n(E[(g(U))^2] - I^2).$$

In summary,

$$E[(\text{sample mean error})^2] = [\int g(x)^2 \, dx - I^2]/n. \qquad (6.13)$$

Regardless of integrand, the expected square error of sample mean Monte Carlo never exceeds that of hit or miss. For if $I \equiv \int g(x) \, dx$

$$E[(\text{hit or miss error})^2] - E[(\text{sample mean error})^2]$$

$$= \text{Var}(n_E/n) - \text{Var}(\bar{g}_n)$$

$$= 1/n[I(1 - I) - (\int g(x)^2 \, dx - I^2)]$$

$$= 1/n(I - \int g(x)^2 \, dx) = 1/n \int g(x)(1 - g(x)) \, dx.$$

But this last term is always non-negative, since for the hit or miss method we must have $0 \le g(x) \le 1$.

Of course equations (6.11) and (6.13) are useless in themselves for estimating error as sampling proceeds because the values I and $\int g(x)^2 \, dx$ are not known. But on the other hand, the law of large numbers allows us to approximate the expected square error: For hit or miss, $n_A/n \to I$ and therefore, for large n,

$$E[(\text{hit or miss error})^2] \simeq (1/n)(n_A/n)(1 - n_A/n). \quad (6.14)$$

Similarly, if $\int g(x)^2 \, dx < \infty$, by the law of large numbers,

$$1/n \sum_{i=1}^{n} (g(U_i))^2 \to \int g(x)^2 \, dx$$

and consequently,

$$E[(\text{sample mean error})^2] = 1/n\{E[g(U)^2] - E[g(U)]^2\}$$

$$\simeq 1/n[1/n \sum_{i=1}^{n} g(x_i)^2 - (\bar{g}_n)^2]. \quad ((6.15)$$

Similar reasoning allows one to conclude that for Algorithm 6.3 (the method for infinite domains) and for large n,

$$E[(\text{sample mean error})^2] \simeq 1/n[1/n \ \Sigma \ (g(x_i)/f(x_i))^2$$

$$- (1/n \ \Sigma \ g(x_i)/f(x_i))^2], \hspace{2cm} (6.15a)$$

the range of summation being, of course, 1 to n.

In the development above, formulas have been obtained for the expected square error of the various elementary Monte Carlo methods related thus far. As we now see, these estimates may in turn be employed to obtain expressions for confidence intervals (see Lindgren, Section 5.3). Thus given any number d, we will see how to compute numbers c_1 and c_2 such that

$$P[c_1 < I < c_2] \geq 1 - d. \hspace{2cm} (6.16)$$

where as usual I denotes $\int g(x) \, dx$. The Chebyshev inequality (Lindgren, p. 120), and the fact that all our Monte Carlo schemes are unbiased lead us to the conclusion that under any Monte Carlo algorithm,

$$P[I_n - e < I < I_n + e] > 1 - \frac{\text{var(error)}}{e^2} \hspace{2cm} (6.17)$$

where I_n is the estimator of I (i.e. n_A/n, $\overline{g}_n$, etc.).

From (6.17) we immediately have the desired bounds by choosing $e = (\text{var(error)}/d)^{1/2}$ and $c_1 \equiv I_n - e$, $c_2 \equiv I_n + e$.

The Chebyshev inequality is known to generally be pessimistic (the interval is needlessly large). If enough samples have been taken that the central limit theorem

(discussed in Section 5.1) is in force, a simple calculation shows that

$$(I_n - I)/V^{1/2} \simeq Y \qquad\qquad (6.18)$$

where Y is the standard normal variable and $V \equiv E[(error)^2]$.
Consequently, if t is a number such that

$$1 - F_Y(t) = P[Y > t] = d/2,$$

then, assuming the approximation in (6.18) is exact, if

$$c_1 = I_n - tV^{1/2} \quad \text{and} \quad c_2 = I_n + tV^{1/2}$$

then (c_1, c_2) is the desired confidence interval, i.e.,
(6.16) holds with equality. We have mentioned earlier
(equations (6.14), (6.15), and (6.15a)) how V may be approximated from the observations $\{g(x_i)\}$.

   We expermentally tested the sample variance and the
central limit theorem confidence interval techniques described in this section on the functions

$$f_1 = 4(1 - x^2)^{1/2} \qquad\qquad f_3 = x^3$$

$$f_2 = \sin(x) \qquad\qquad f_4 = e^x$$

which have been subjects in earlier computational experiments. Specifically, we computed the theoretical standard
deviation (defined to be the square root of the variance)

for various sample sizes of hit or miss and sample mean
estimates by using the variance formulae (6.11) and (6.13).
Then, simulation runs were made and the sample standard
deviations of hit or miss and sample mean rules were calculated by taking the square roots of the expressions (6.14)
and (6.15) respectively. The results of these calculations
are presented in Table 6.5. From this experiment, we may
develop some feeling for the accuracy of the sample estimate
of the Monte Carlo error.

One may verify from a standard normal distribution
table (such as Table I of Lindgren) that the probability
that the error is less than one standard deviation is
0.6826. Thus if the central limit approximation (6.18) is
presumed to hold exactly, then P[error $\leq$ standard deviation]
= 0.6826. Thus in light of our discussion of confidence
intervals, we would expect that 68% of the time, the error
is less than the standard deviation. That is, 68% of the
time, I ε [I_n - σ, I_n + σ], where σ denotes the standard
deviation. In Table 6.5, we have printed out the errors
and the reader may confirm that in most cases, the error
does turn out to be less than the standard deviation. Note
also that the errors and standard deviations of the sample
mean integration rule are consistently* less than the hit
or miss errors and standard deviations, illustrating the
general results at the beginning of this section.

---

*In five entries, the sample mean error does exceed the hit
or miss error.

Table 6.5

Comparison of Errors and Standard Deviations For Hit or Miss and Sample Mean Methods

$$\int_0^1 4\sqrt{1-x^2}\,dx = 3.14159$$

N	Errors		Sample Standard Deviations		Theoretical Standard Deviations	
	Hit or Miss	Sample Mean	Hit or Miss	Sample Mean	Hit or Miss	Sample Mean
20	.342	.113	.420	.217	.367	.200
50	.102	.183E-1	.244	.139	.232	.126
500	.184E-1	.150E-1	.729E-1	.409E-1	.734E-1	.399E-1
10000	.244E-1	.695E-2	.166E-1	.902E-2	.164E-1	.893E-2

$$\int_0^1 \sin(x)\,dx = .45970$$

N	Errors		Sample Standard Deviations		Theoretical Standard Deviations	
	Hit or Miss	Sample Mean	Hit or Miss	Sample Mean	Hit or Miss	Sample Mean
20	.597E-1	.548E-1	.112	.579E-1	.111	.554E-1
50	.403E-1	.199E-1	.714E-1	.358E-1	.705E-1	.350E-1
500	.370E-2	.851E-2	.223E-1	.112E-1	.223E-1	.111E-1
10000	.420E-2	.912E-3	.498E-2	.248E-2	.498E-2	.248E-2

Table 6.5 Continued

$$\int_1^0 x^3\, dx = .25000$$

N	Errors		Sample Standard Deviations		Theoretical Standard Deviations	
	Hit or Miss	Sample Mean	Hit or Miss	Sample Mean	Hit or Miss	Sample Mean
20	.100	.334E-1	.819E-1	.711E-1	.968E-1	.634E-1
50	.500E-1	.635E-2	.655E-1	.448E-1	.612E-1	.401E-1
500	.400E-2	.574E-2	.193E-1	.129E-1	.193E-1	.127E-1
10000	.670E-2	.178E-2	.429E-2	.285E-2	.433E-2	.285E-2

$$\int_0^1 e^x\, dx = 1.71828$$

N	Errors		Sample Standard Deviations		Theoretical Standard Deviations	
	Hit or Miss	Sample Mean	Hit or Miss	Sample Mean	Hit or Miss	Sample Mean
20	.486E-1	.924E-1	.297	.119	.293	.110
50	.130	.299E-1	.181	.739E-1	.185	.696E-1
500	.112E-1	.147E-1	.588E-1	.222E-1	.586E-1	.220E-1
10000	.420E-3	.239E-2	.131E-1	.493E-2	.131E-1	.492E-2

6.5 EFFICIENT MONTE CARLO INTEGRATION

It turns out that the Monte Carlo methods of hit or
miss and sample mean, while highly appealing for their
directness and simplicity, are inferior to certain more
sophisticated techniques because the latter achieve significantly smaller error. The hit or miss method was exceed-
ingly popular in the early days of Monte Carlo and, because
of its inefficiency, it gave the Monte Carlo method a bad
image as being grossly inferior to classical integration
(such as described, for example, in Henrici (1964), Chapter
13). In this connection, Hammersley and Handscomb (1964)
write (p. 9)

> In the last few years Monte Carlo methods
> have come back into favor. This is mainly
> due to better recognition of those problems
> in which it is the best, and sometimes the
> only, available technique. Such problems
> have grown in number, partly because improved
> variance-reducing techniques just discovered
> have made Monte Carlo efficient where it had
> been previously inefficient....

Control variate method. We will now discuss a certain
modern small-variance Monte Carlo integration method known
as control variate Monte Carlo. It has the strengths representative of sophisticated new Monte Carlo algorithms in
that it is generally highly successful in reducing the
error. And it shares weaknesses typical of advanced techniques: One must do some planning to employ it effectively.
In contrast to the sample mean and hit or miss methods, in
control variate integration, procedural decisions must be
made before the method can be coded.

Let $g(x)$ be the integrand and, for now, assume the domain of integration is the unit interval. The idea of control variate Monte Carlo is that the engineer seeks some function $h(x)$ which approximates $g(x)$ well and has the further important property that it is easy to integrate (e.g., a polynomial). Then

$$\int g(x)\ dx = H + \int (g(x) - h(x))\ dx$$

where H denotes the easily-found integral of $h(x)$. The function $h(x)$ is called the control variate. Now we apply sample mean Monte Carlo to the function $g(x) - h(x)$. If U denotes the uniform variable, then as in (6.13), the error of estimation based on Monte Carlo observations is

$$E[(error)^2] = 1/n[E[(g(U) - h(U))^2]$$

$$- [\int (g(x) - h(x))\ dx]^2].$$

As in the case of the sample mean method, $E[(error)^2]$ can be approximated by the sample variance of the sequence $\{g(u_i) - h(u_i)\}$.

The control variate method throws some problem-solving burden on the user. Given the integrand $g(x)$, the user must somehow provide the control variate approximation $h(x)$, and there is no definitive science for this activity, although various numerical analysis subjects are extremely relevant. One appealing and universally applicable method for finding a control variate is to select a few points $\{g(x_i)\}$ perhaps at random and let $h(x)$ be an interpolation polynomial (see Henrici, 1964, Chapter 9) for these points. Under weak

conditions, interpolation polynomials converge uniformly
with increasingly many interpolation points to the target
function.

In a computer experiment we performed, we adopted an
even simpler, but less effective, procedure. Twenty domain
points $\{x_i\}_{i=1}^{20}$ are selected at random. We define $h(x)$ to
be the step function $h(x) = g(x_j)$ where j is chosen so that

$$|x - x_j| \leq |x - x_k|, \ 1 \leq k \leq 20.$$

For purposes of comparison, this control variate Monte
Carlo scheme was applied to the same integrands f_1, f_2, f_3,
f_4 as preceding studies. In Table 6.6, we have presented
the results of our computational experiment.

The errors of simulation by sample mean and the control
variate method as well as the sample standard deviations
are tabulated for comparison. The vastly improved accuracy
of the control variate algorithm is evident.

If the user desires to integrate improper integrals,
Algorithm 6.3 may be applied to the residual integrand
$g(x) - h(x)$. The reader will readily see that control
variate Monte Carlo may be used without modification for
integrating functions of several variables. In fact, the
step function technique for constructing control variates
is applicable in higher dimension spaces.

Control variate Monte Carlo is an excellent illustration of a maxim repeated in most of the treatises on Monte
Carlo, to wit: in preparing a problem for Monte Carlo
analysis one should, as much as possible, factor out the
deterministic part of the problem. The control variate
technique calls for integrating exactly a function which is

Table 6.6

Comparison of Control Variate and Sample Mean Methods

$$\int_0^1 4\sqrt{1-x^2}\,dx = 3.14159$$

N	Control Variate Approximation	Errors		Sample Standard Deviations	
		Control Variate	Sample Mean	Control Variate	Sample Mean
20	3.0541	.875E-1	.113	.229E-1	.217
50	3.1490	.741E-2	.183E-1	.387E-1	.139
500	3.1338	.779E-2	.150E-1	.130E-1	.409
10000	3.1416	.434E-4	.695E-2	.290E-2	.902E-2

$$\int_0^1 \sin(x)\,dx = .45970$$

N	Control Variate Approximation	Errors		Sample Standard Deviations	
		Control Variate	Sample Mean	Control Variate	Sample Mean
20	.4714	.117E-1	.548E-1	.691E-2	.579E-1
50	.4563	.340E-2	.199E-1	.610E-2	.358E-1
500	.4596	.523E-4	.851E-2	.231E-2	.112E-1
10000	.4596	.138E-3	.912E-3	.516E-3	.248E-2

Table 6.6 Continued

$\int_0^1 x^3\, dx = .25000$

N	Control Variate Approximation	Errors		Sample Standard Deviations	
		Control Variate	Sample Mean	Control Variate	Sample Mean
20	.2804	.304E-1	.334E-1	.713E-2	.711E-1
50	.2460	.403E-2	.635E-2	.128E-2	.448E-1
500	.2518	.179E-2	.574E-2	.433E-2	.129E-1
10000	.2496	.377E-3	.178E-2	.974E-3	.285E-2

$\int_0^1 e^x\, dx = 1.71828$

N	Control Variate Approximation	Errors		Sample Standard Deviations	
		Control Variate	Sample Mean	Control Variate	Sample Mean
20	1.7540	.356E-1	.924E-1	.133E-1	.119
50	1.7108	.751E-2	.299E-1	.163E-1	.739E-1
500	1.7195	.125E-2	.147E-1	.581E-2	.222E-1
10000	1.7178	.440E-3	.239E-2	.130E-2	.493E-2

as close as possible to the difficult integrand. The residual function $g(x) - h(x)$ then has greatly reduced magnitude (compared to $g(x)$ itself) and the error is correspond-
ingly reduced.

The antithetic variate method.

The antithetic variate
method draws its name from the insight that if U denotes a
random number and $t_1(U)$ and $t_2(U)$ denote two unbiased estimates of some integral I, and if we take the sample average
$t_3(U) \equiv 1/2(t_1(U) + t_2(U))$, then one may hope that $\text{var}(t_3(U))$
is much less than either the variance of $t_1(U)$ or $t_2(U)$.
This can occur if $t_1(U)$ and $t_2(U)$ are negatively correlated.
For in that case, we have

$$\text{var}(t_3(U)) = 1/4(\text{var}(t_1(U)) + \text{var}(t_2(U))$$
$$+ 2\text{covar}(t_1(U), t_2(U)). \qquad (6.19)$$

The last term in (6.19) is the covariance (defined in
Lindgren, p. 122) of $t_1(U)$ and $t_2(U)$. If $t_1(U)$ and $t_2(U)$
have a negative correlation coefficient, the covariance will
be negative, and this gives the mechanism whereby the antithetic variate method can be extremely effective in reducing
the estimation variance (or what is the same for unbiased
estimates, the expected square error) of the estimator. As
an example of how one might construct an antithetic variate
which reduces the error below that of sample mean Monte
Carlo, suppose it is known that $g(x)$ is a monotonically
increasing function. If u is a random number near 0, then
loosely speaking $g(u)$ is probably smaller than I, and corre-
spondingly, if u is near 1, $g(u)$ will in all likelihood be
too large. One can hope to cancel the errors by defining a

new estimate $1/2[g(u) + g(1 - u)]$. If $g(u)$ is too small
then $g(1 - u)$ will have a good chance of being too large,
and conversely. This example contains the essence of the
antithetic variable approach, but the literature is extensive and the methods and results are relatively sophisticated. We refer the reader to the fundamental papers of
Hammersley and Morton (1956), Tukey (1957), Halton and
Handscomb (1957); results of these works are summarized in
Section 5.6 of Hammersley and Handscomb (1964).

It is possible to rephrase the aim in constructing an
antithetic variable. One seeks an unbiased estimator $t(U)$
of I, such that when $t(\cdot)$ is regarded as a function of a
real variable x, the quantity $M \equiv \max_{0 \leq x_1 \leq x_2 \leq 1} |t(x_1) - t(x_2))|$
is as small as possible. For it is evident that
$\text{var}(t(U)) \leq M^2$. How to achieve this aim remains a problem
for the Monte Carlo user. One simple approach is the
following. Assume $g(x)$ is a function having a continuous
first derivative bounded in magnitude on the interval $[0,1]$
by some number C. Then one may quickly verify that if U
is a random number, the estimator

$$t(U) = 1/m \sum_{i=0}^{m-1} g((i + U)/m) \qquad (6.20)$$

is an unbiased estimator for I and further, for any u,
$|t(u) - I| \leq C/m$.

In Table 6.7, we present the results of applying the
estimator (6.20) to our usual four functions. We took
$m = 20$, in all cases.

While it is true that the antithetic variate theory is
a mature branch of Monte Carlo integration technology, its

Table 6.7

Antithetic Variate Monte Carlo Integration

(Note: N is the number of evaluations of the function).

$$\int_0^1 4\sqrt{1 - x^2}\ dx = 3.14159$$

N	Approximation	Error	Sample Standard Deviation
20	3.0464	.952E-1	.231
40	3.1101	.315E-1	.149
500	3.1400	.161E-2	.401E-1
10000	3.1432	.163E-2	.891E-2

$$\int_0^1 \sin(x)\ dx = .45970$$

N	Approximation	Error	Sample Standard Deviation
20	.4791	.194E-1	.560E-1
40	.4660	.631E-2	.395E-1
500	.4599	.187E-3	.111E-1
10000	.4593	.398E-3	.248E-2

Table 6.7 Continued

$$\int_0^1 x^3 \, dx = .25000$$

N	Approximation	Error	Sample Standard Deviation
20	.2736	.236E-1	.696E-1
40	.2576	.766E-2	.465E-1
500	.2503	.274E-3	.127E-1
10000	.2495	.468E-3	.283E-2

$$\int_0^1 e^x \, dx = 1.71828$$

N	Approximation	Error	Sample Standard Deviation
20	1.7583	.400E-1	.115
40	1.7312	.130E-1	.794E-1
500	1.7187	.420E-3	.220E-1
10000	1.7175	.809E-3	.492E-2

extension to integrands of several variables has proven
difficult; and yet multivariate integration is the activity
within numerical integration in which the Monte Carlo
approach seems the most fruitful. These matters will be
discussed in the section to follow.

6.6   WHEN IS MONTE CARLO THE BEST INTEGRATION METHOD?

The question asked in the title of this section is one
of considerable importance and the center of a good deal of
controversy in the numerical analysis literature. It seems
impossible to provide a definitive answer at the present
stage of development, but some analytic results are related
here which are cogent to the issue.

The importance of having good numerical integration
schemes available is evident to almost anybody who has
worked at an advanced level in data analysis. This author's
experience has led him to the conclusion that one of the
central limitations to practical implementation of decision
theory and sequential decision theory is the difficult
multivariate integrations that must be done to find Bayes
risk. In the monograph (Yakowitz, 1969), it is seen that
one must stick to very simple problems which do not require
numerical integration in order to avoid being swamped by
computer expense in stochastic dynamic programming. The
paper (Yakowitz, et al, 1975) shows that numerical Bayes
analysis of even the very popular gamma random variable
requires numerical trickery.

For well-behaved real functions, quadrature methods
such as discussed in elementary textbooks (e.g. Henrici,

1964, Chapter 13) are available in most computer center
libraries and are extremely accurate. Monte Carlo methods
are not competitive.

But if the function fails to be extremely regular (i.e.
to have continuous derivative of moderate order), the standard textbook methods are less attractive. For example,
consider the continuous function (chosen in an arbitrary
fashion):

$(5/3)x,$ $x \in [0, 0.15]$

$-x + 0.4,$ $x \in [0.15, 0.2]$

$0.5x + 0.1,$ $x \in [0.2, 0.4]$

$-2x + 1.1,$ $x \in [0.4, 0.45]$

$8/3x - 1,$ $x \in [0.45, 0.6]$

$-1/2x + 0.9,$ $x \in [0.6, 0.7]$

$2.5x - 1.2,$ $x \in [0.7, 0.8]$

$-x + 1.6,$ $x \in [0.8, 0.9]$

$3x - 2,$ $x \in [0.9, 1.00]$.

In a crude numerical experiment, we integrated the
above function by the control-variate Monte Carlo method
described in Section 6.5 as well as by 16-point Gauss quadrature, one of the most popular and theoretically most attractive classical numerical integration methods. The Monte
Carlo algorithm, which used 10,000 iterations, gave less
error by an order of magnitude. The error by Gauss quadrature was 0.077, and by Monte Carlo, 0.005.

The classical deterministic integration methods are
not immediately extensible for integrating over higher
dimension space, and in multivariate integration, therefore,
the situation is much more challenging. At the University
of Arizona Computer Center, for example, no packaged programs are available for multivariate integration. Monte
Carlo methods do not suffer so greatly in generalization to
higher dimension. The "Hit or Miss," "Sample Mean," and
"control variate" algorithms and bounds are directly applicable with the understanding that the random quantities u_i
are vector-valued random variables uniformly distributed
on the unit hypercube. Some authors imply that Monte Carlo
is the method of choice for multivariate integration. The
treatise Approximate Calculation of Multiple Integrals by
A. A. Stroud (1971) takes a less enthusiastic view of Monte
Carlo. In Chapter 6, Stroud describes Monte Carlo-related
techniques called number-theoretic methods (due to the
Soviet mathematician Korobov) which achieve an error bound
for the difference E_N between the integral and its estimate
based on N sample points, namely

$$E_N \leq C(\ln(N))^a/N. \qquad (6.21)$$

The only restriction on the integrand f is that $\frac{\partial^n f}{\partial x_1 \ldots \partial x_n}$
be continuous. The measure of quality corresponding to
E_N in Monte Carlo theory is the standard deviation σ_N,
which can invariably be written in the form

$$\sigma_N = C'/\sqrt{N}. \qquad (6.22)$$

It is evident that asymptotically, at least, the "number theoretic" method results in smaller error (but it should be borne in mind that the calculations required to implement the number theoretic method are more difficult). A method due to Hlawka (1961) also achieves the order of convergence of equation (6.19), according to (Zaremba (1968)). Zaremba (1968) gives supplementary information on the mathematics of Monte Carlo integration, and advances the interesting notion that randomness is not what makes Monte Carlo work relatively well for multivariate integration, but the equidistribution characteristics. Further, the thrust of his paper is to show that, regardless of the dimension of the domain of integration, better error bounds can be achieved by choosing the u_i's "wisely," in the sample mean formula $1/n \sum_{i=1}^{n} g(u_i)$, instead of randomly.

Formulae which exploit this viewpoint are termed by Stroud (1971) and others quasi-Monte Carlo methods, and the number theoretic methods of Korobov, Hlawka (1961), and Davis and Rabinowitz (1956) fall into this category.

In a master's thesis, Mason (1976) carefully compared various Gaussian and quasi-Monte Carlo techniques on integration problems involving three to six variables. The experimental evidence assembled seems to point to the conclusion that for well-behaved functions, classical Gaussian quadrature is quite competitive with quasi-Monte Carlo procedures. But if the functions are discontinuous or even fail to possess continuous first partial derivatives everywhere, Monte Carlo and quasi-Monte Carlo methods are competitive and often superior. Chapter 6 of Stroud reports computer studies which also are consistent with this conclusion.

In Figure 6.2, we have reproduced a diagram from Mason (1976) which compares as a function of processing time the estimates of the integral

$$\int_E [(\sum_{i=1}^{4} |x_i - .5|)^5] \, dx,$$

where E is the four dimensional unit hypercube. The stratified sampling scheme is a pure Monte Carlo method discussed in Hammersley and Handscomb (1964). The Korobov and Davis-Rabinowitz algorithms are quasi-Monte Carlo methods described in Stroud (1971). The Gauss-Legendre product formula is a classical quadrature technique. One sees from this figure that, whereas the classical Gauss-Legendre estimate is fluctuating erratically as computation effort increases, the Monte Carlo and quasi-Monte Carlo method are converging toward a common value. Because it would constitute too much of a digression here, we refer the reader to Stroud (1971) for particulars of the definitions and constructions of the quasi-Monte Carlo schemes used in the study. Stroud (1971) also reports computer simulation results that accord well with our assessment of the relative merits of classical and Monte Carlo methods.

The mathematical theory of multivariate integration is still in an early state, and for the engineer with the immediate integration need, the Monte Carlo and quasi-Monte Carlo methods are attractive for their simplicity. Some computations and theoretic developments in Stroud (1971, Chapter 6) suggest that even for extremely smooth functions, the Monte Carlo approach has effectiveness comparable to n-dimensional generalizations of standard quadrature formulae.

Stroud (1968, 1969) and Franke (1971) report progress toward
extending Gauss quadrature theory to multiple integration.

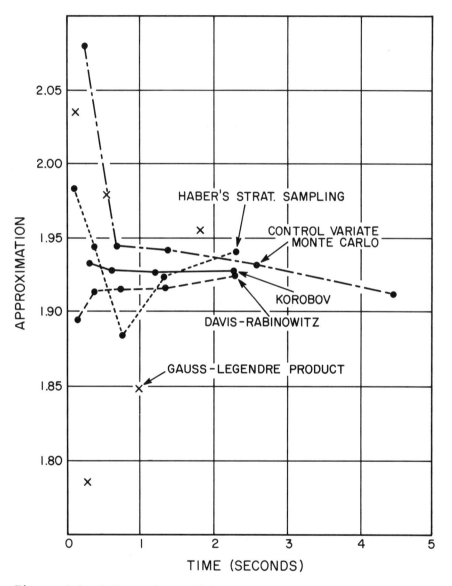

Figure 6.2   A Comparison of Several Multivariate Integration
Methods

6.7   MONTE CARLO SOLUTION OF BOUNDARY-VALUE PROBLEMS

The Monte Carlo method has application to solving
differential equation systems, and techniques for solving
boundary-value problems involving Laplace's equation,
Poisson's equation, the heat equation, and second order
linear elliptic equations are described in Shreider (1966,
Chapter 1). In fact, differential equations can often be
viewed as models for microscopically random physical phenomena, and consequently simulation is, to some extent, a return to reality, rather than a second-order abstraction from
reality, as are other numerical techniques. The name "Monte
Carlo" was first dubbed to a random method for analyzing
otherwise numerically unsolvable problems arising in the
study of neutron transport, design of nuclear reactors, and
other problems in particle physics at Los Alamos. The first
paper, and a very readable one, using the appellation "Monte
Carlo" was by Metropolis and Ulam (1949), but they credit
the name to the mathematician J. von Neumann. The problems
in nuclear engineering which motivated early developments
in the Monte Carlo method as well as the techniques themselves are described in the book by Spanier and Gelbard
(1969).

In this section, we will illustrate the Monte Carlo
method for solving differential equations by solving a two-
dimensional boundary value problem involving Laplace's equation (otherwise known as the Dirichlet problem; see, for
example, W. Kaplan, Advanced Calculus, 1952, pp. 591-599).
Suppose we have a simple closed curve C in the plane on
which is defined a continuous function f(s). Complex
variable theory assures us of the existence of a unique

function u(s) which is harmonic (satisfies Laplace's equa-
tion: $\nabla u = \dfrac{\partial^2 u}{\partial x^2} + \dfrac{\partial^2 u}{\partial y^2} = 0$) in the interior D of C, is equal
to f on C, and is continuous everywhere. The method to
follow will allow us to approximate the value of this function at any point in D. This problem has considerable physical motivation. For example, if f is a voltage (static
electricity) or a velocity potential (fluid dynamics) on the
curve C, then u will be, respectively, the voltage or potential in the region D.

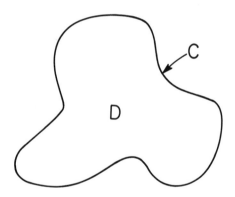

Figure 6.3   A Simple Closed Curve

In the fashion of the classical procedures of numerical
analysis, we begin by establishing a grid of points over
the region D as illustrated in Figure 6.4. The points interior to C are represented by dots, and the points considered to lie on the boundary C are illustrated by crosses,

and if they do not actually lie on C, a value h(s') is assigned to them which is approximately equal to the value that f takes on at the point on C closest to s'.

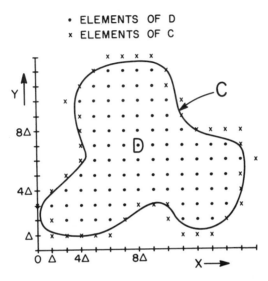

Figure 6.4   Grid Associated with a Region

The usual basis for numerical methods for solving differential equations is the assumption that the grid width Δ so fine that the differential equations are accurately approximated by difference equations:

$$\frac{\partial}{\partial x} u(x, y) \simeq (u(x + \Delta, y) - u(x, y))/\Delta,$$

$$\frac{\partial^2}{\partial x^2} u(x, y) \simeq \left[\frac{\partial u}{\partial x}(x, y) - \frac{\partial u}{\partial x}(x - \Delta, y) \right] \frac{1}{\Delta} \qquad (6.23)$$

$$\simeq [u(x + \Delta, y) - u(x, y)] - [u(x, y)$$

$$- u(x - \Delta, y)]/\Delta^2$$

$$\simeq [u(x + \Delta, y) + u(x - \Delta, y) - 2u(x, y)]/\Delta^2.$$

Similarly,

$$\frac{\partial^2}{\partial y^2} u(x, y) \simeq \frac{[u(x, y + \Delta) + u(x, y - \Delta) - 2u(x, y)]}{\Delta^2}.$$

Then as a difference equation, Laplace's equation becomes

$$0 = \frac{\partial^2 u}{\partial x^2} + \frac{\partial^2 u}{\partial y^2} \simeq [u(x + \Delta, y) + u(x - \Delta, y)$$

$$+ u(x, y + \Delta) + u(x, y - \Delta) - 4u(x, y)]/\Delta^2 \quad (6.24)$$

or solving for $u(x, y)$

$$u(x, y) = 1/4 [u(x + \Delta, y) + u(x - \Delta, y) + u(x, y + \Delta)$$

$$+ u(x, y - \Delta)] \quad\quad\quad\quad\quad\quad (6.25)$$

and u is harmonic if and only if it satisfies (6.25). If
the probability of moving from (x, y) to each of the points
$(x + \Delta, y)$, $(x - \Delta, y)$, $(x, y + \Delta)$, $(x, y - \Delta)$ is 1/4, then
(6.25) admits a probabilistic interpretation, namely the
expected return associated with being at (x, y) in terms of
the expected return associated with being at any of the
neighboring points. We take the reward associated with a
random walk that reaches boundary point (x', y') to be

f(x', y'). Thus the boundary is viewed as an absorption
barrier for a two-dimensional random walk starting at point
(x, y). Letting $X'Y'|(x, y)$ be the random variable denoting
the boundary point absorbing a walk starting at (x, y), we
have

$$u(x, y) = E[f(X', Y')|(x, y)] \qquad\qquad (6.26)$$

and (6.25) is satisfied if successive steps are chosen from
the four grid points neighboring (x, y) by a law assigning
equal probability to each of these points. Also for every
boundary point (x', y'),

$$u(x', y') = E[f(X', Y')|(x', y')] = f(x', y').$$

u(x, y) is smooth in the sense that it averages the neighboring grid points. In summary, u(x, y) is harmonic in D,
agrees with f on C, and is "continuous" on C $\cup$ D and thus is
the unique solution to the boundary value problem. u(x, y)
is approximated, of course, by repeatedly performing random
walks starting at (x, y) and recording for each walk i,
i = 1,2,...,N, the value $f(x_i', y_i')$ that f takes on at the
boundary point (x_i', y_i') terminating the ith random walk.
(A consequence of probability theory is that absorption is
certain to occur.) By the law of large numbers (Theorem
5.1) with a probability of 1, $1/N \sum_{i=1}^{N} f(x_i', y_i') \to u(x, y)$
as $N \to \infty$.

The Monte Carlo technique described above for the solution of the Dirichlet problem is readily modified for application to other elliptic partial differential equations,
including nonhomogeneous ones.

Monte Carlo methods are particularly attractive if the solution u(x, y) is required at only one domain point. Classical finite difference schemes for partial differential equations usually require the solution at every point as an intermediate step to getting the solution at a desired point. We saw the same phenomenon of the Monte Carlo method's unique capability of efficiently finding single coordinates of the solution vector in our study of linear equations in Chapter 3. Usually, classical numerical methods proceed by converting linear partial differential equations into linear algebraic equations with one variable for each solution coordinate. We refer the reader to Isaacson and Keller (1966, Chapter 9), for a summary of conventional approaches.

Program 6.1 gives a Monte Carlo approximation to the boundary value problem in which C is a unit circle, the starting point $(x, y) = (1/2, 0)$, and $f(x, y) = e^x \cos(y)$ (= real part of e^{x+iy}). Δ is taken to be 0.05, and we are able to get by with very little use of memory by, at each iteration in the random walk, determining whether the current point (x, y) is a boundary point by computing $x^2 + y^2$ and comparing it with 1. If $x^2 + y^2 \geq 1$, then (x, y) is a boundary point and $f(x, y) = e^x \cos(y)$ is the value of that walk.

We have chosen our problem in such a way as to be able to check our estimate easily. As f(x, y) is the real part of an analytic function, the problem has a solution given by Poisson's formula (Kaplan, 1952, pp. 546-547), and consequently we will be able to check the accuracy of our Monte Carlo answer. Transforming u(x, y) to a function u(r, θ) of polar coordinates, Poisson's formula is

$$u(r, \phi) = \frac{1}{2\pi} \int_0^{2\pi} \frac{R^2 - r^2}{R^2 - 2rR \cos(\theta - \phi) + r^2} u(R, \theta) \, d\theta.$$

R is the radius of C, r the modulus of the random walk starting point (x, y), and ϕ is the polar angle of (x, y). Thus, for our problem,

$$u(\tfrac{1}{2}, 0) = \frac{1}{2} \int_0^2 \left[\frac{3/4\{e^{\cos(\theta)} \cos(\sin(\theta))\}}{5/4 - \cos(\theta)} \right] d\theta. \qquad (6.27)$$

We have employed (6.27) to compute that to four significant figures, $u(\tfrac{1}{2}, 0) = 1.649$. One sees that the relative error of the random walk solution was about 1%, after 500 iterations.

Program 6.1

Monte Carlo Solution to a Boundary-Value Problem

```
      PROGRAM BNDVAL(OUTPUT,TAPE6=OUTPUT)
C         BNDVAL FINDS THE SOLUTION TO A BOUNDRY VALUE
C         PROBLEM AT A POINT BY THE MONTE CARLO METHOD
C
C         U,V- FIND THE SOLUTION AT THIS POINT
C         IWALK- COUNTS RANDOM WALKS
C         APPROX- APPROXIMATE SOLUTION
C
      WRITE(6,80)
80    FORMAT(1H1,"RANDOM WALK APPROXIMATION TO"/
     1       1X,"THE BOUNDRY VALUE PROBLEM"//
     2       1X,"NUMBER OF"/
     3       1X,"WALKS        APPROXIMATION"/)
      DATA U,V / 0.5, 0.0 /
      SUM = 0.0
      DO 100 IWALK=1,2000
        X = U
        Y = V
C         "WALK" TILL ABSORBED
50      X = X + SIGN(0.05, RANF(DUMMY) - 0.5 )
        Y = Y + SIGN(0.05, RANF(DUMMY) - 0.5 )
        IF( X**2 + Y**2 .LT. 1.0 ) GOTO 50
```

```
      SUM = SUM + EXP(X)*COS(Y)
      IF( MOD(IWALK, 100) .NE. 0.0 ) GOTO 100
       APPROX = SUM/FLOAT(IWALK)
       WRITE(6,81) IWALK,APPROX
81     FORMAT(1X,I5,6X,G12.5)
100    CONTINUE

   STOP
   END
```

RANDOM WALK APPROXIMATION TO
THE BOUNDRY VALUE PROBLEM

NUMBER OF WALKS	APPROXIMATION
100	1.6071
200	1.6428
300	1.6334
400	1.6500
500	1.6241
600	1.6443
700	1.6325
800	1.6362
900	1.6528
1000	1.6611
1100	1.6440
1200	1.6342
1300	1.6342
1400	1.6399
1500	1.6394
1600	1.6424
1700	1.6456
1800	1.6412
1900	1.6461
2000	1.6536

The reader interested in further discussion of the
connection between random processes and potential theory
may wish to review the Scientific American article "Brownian
Motion and Potential Theory," by Hersh and Griego (1969),
which credits the Japanese probabilist Kakutani with recognizing the connection between the Dirichlet problem and
random walks (or rather Brownian motion, which is the
limiting process of random walks).* The Monte Carlo method
of this section depends on this insight.

---

*Another Scientific American article related to this Chap-
ter is entitled "The Monte Carlo Method" by D. McCracken
(1955).

APPENDIX

PROBABILITY EXPERIMENTS INDUCED BY FINITE MEASURES

Let μ be a totally finite measure (signed or otherwise)
defined on a σ-field G of subset's of an abstract set Ω.
The measure μ admits a unique Jordan decomposition (See
Halmos (1950), p. 123) $\mu(E) = \mu_1(E) - \mu_2(E)$, $E \in G$, where μ_1
and μ_2 are both non-negative totally finite measures. Thus
if $a = \mu_1(\Omega)$ and $b = \mu_2(\Omega)$, then $P_1 = \mu_1/a$ and $P_2 = \mu_2/b$
are probability functions on the field of events G. Let
Z_1, Z_2 denote the probability experiments (Ω, G, P_1),
(Ω, G, P_2) respectively and suppose $g(x)$ is any G-measurable
real-valued function defined on Ω. $\{x_i\}_{i=1}^n$ and $\{y_i\}_{i=1}^n$
denote samples of size n on Z_1 and Z_2 respectively. By
the law of large numbers, if $g(x)$ is μ-integrable, the sample mean $(a/n \sum_{i=1}^n g(x_i) - (b/n) \sum_{i=1}^n g(y_i))$ converges in
probability to

$$E_X[g(X)] - E_Y[g(Y)] = \int g(x)\, d\mu_1 - \int g(x)\, d\mu_2$$

$$= \int g(x)(d\mu_1 - d\mu_2) = \int g(x)\, d\mu.$$

EXERCISES

1) Compare the errors and sample standard deviations of
    hit or miss and sample mean Monte Carlo for some inte-
    grand of your choosing. (It should be a function whose
    integral can be found exactly.)

2) Use Lagrange interpolation (Henrici, 1964, p. 184, for
    example) to find a "control" function for $\sin^2(\pi x)$,
    $0 \leq x \leq 1$. Let the interpolation points be $x = 0, 0.2$,
    $0.4, \ldots, 1.0$. With this polynomial as control function,
    compare the performance of the control variate technique
    with that of sample mean Monte Carlo for computing
    $\int_0^1 \sin^2(\pi x)\, dx$. Make a similar comparison for some
    function with discontinuous first derivative.

3) Describe how to generalize the "step function" control
    function technique used in Section 6.4 to higher dimen-
    sion spaces. Use your technique to give control variate
    Monte Carlo estimates for the four dimensional integral

$$\int_E [\sum_{i=1}^4 |x_i - 0.5|]^5 \, dx$$

described in connection with Figure 6.4. Through simulation, compare the sample standard deviation of the
control variate estimator with that of sample mean for
this problem.

4) A "product" quadrature formula for integration over the
unit hypercube is obtained by using a classical one-
dimensional quadrature formula over each dimension of
the cube. Thus if w_i's are the weights and x_i's are
the quadrature points of a one dimensional formula,
then in the square E_2, $\int_{E_2} f(x) \, dx$ is approximated by
$\sum_{i,j} w_i w_j f(x_i, x_j)$. Use some such formula on the integrand in Problem 3 and compare the accuracy of using a
10^4 point product rule (10 points in each dimension)
with a sample mean estimate using 10^4 points.

5) Explain a reasonably efficient way to generate uniform
random samples over S_n, the interior of a unit sphere
in n dimensions. Use your rule to obtain a sample-mean
Monte Carlo estimate for

$$\int_{S_3} \exp(xyz^2) \, dx \, dy \, dz.$$

According to (Stroud, 1971, p. 36) the value of this
integral is, to five decimal places, given by 4.190604.

6) Let $I = \int_0^1 f(x) \, dx$ and I_n denote an n-sample Monte Carlo
estimate of I. In the methods discussed in this chapter,

$$E[(I - I_n)^2] = C/n.$$

Through the control variate and antithetic variate
techniques, we were able to reduce the magnitude of the
scaling constant C, but the mean square error still is
inversely proportional to n. It is possible to vastly
accelerate the rate of convergence by suitable scaling
of the observed functional values. Let $\{U_i\}_{i=1}^n$ denote
n random numbers and for i = 1,...,n, w_i denotes the
length of the interval of points in [0, 1] which lie
closer to U_i than to the other random numbers. Our
weighted Monte Carlo estimate of I is given by

$$Q_n = \sum_{i=1}^n w_i f(U_i).$$

It is possible to show that if f has a continuous
second derivative in [0, 1], then

$$E[(Q_n - I)^2] = C/n^4.$$

a) For various values of n, try this estimator out on
    the five functions studied in Table 6.2. Experi-
    mentally study the sample variance and see that it
    decreases approximately as $1/n^4$.

b) See if you can prove the asserted convergence rate.
    This is not an easy exercise. Make an ordered sam-
    ple of the U_i's and then view the weighted Monte
    Carlo formula as a trapezoidal rule and use error

bounds from classical quadrature theory. If $\{U_{(j)}\}_{j=1}^{n}$ denotes the ordered sample, then the random vector $\{U_{(i)} - U_{(i-1)}\}$ obeys the Dirichlet law.*

7) Give a Monte Carlo method (akin to the one presented for solving the Dirichlet problem) for approximating the solution to the second order linear homogeneous ordinary differential equation

$$au_{xx} + bu_{x} + cu = 0, \; u(0) = c_1, \; u(1) = c_2.$$

In the above, a, b, c, c_1, and c_2 are constants and u is a function of the real variable x on [0, 1]. Explain what modifications of your method are necessary if there is a forcing function or if a, b, and c depend on x.

8) Show that the Monte Carlo solution of the Dirichlet problem can be viewed, after our discretization of the function domain has been accomplished, as per Figure 6.4, as a linear algebraic equation. Compare our solution of the Dirichlet problem with the Markov chain method for solving linear algebraic equations which was developed in Chapter 3.

---

*The Dirichlet variate (not to be confused with the Dirichlet problem) is described, for example, in (Wilks, 1962, pp. 177-178).

REFERENCES FOR CHAPTER 6

Davis, P., and P. Rabinowitz, (1956). "Some Monte Carlo
    Experiments in Computing Multiple Integrals." Mathe-
    matics of Computation, 10, pp. 1-8.

Franke, R., (1971). "Orthogonal Polynomial and Approximate
    Multiple Integration." SIAM J. Numer. Anal., 8,
    pp. 757-766.

Halmos, P., (1950). Measure Theory. van Nostrand, Prince-
    ton, New Jersey.

Halton, J., and D. Handscomb, (1957). " A Method for In-
    creasing the Efficiency of Monte Carlo Integration."
    J. Assoc. Comp. Mach., 4, pp. 329-340.

Hammersley, J., and D. Handscomb, (1964). Monte Carlo
    Methods. Methuen, London.

Hammersley, J., and K. Morton, (1956). "New Monte Carlo
    Technique: Antithetic Variates." Proc. Cambridge
    Phil. Soc., 52, pp. 449-475.

Handscomb, D., (1958). "Proof of the Antithetic Variates
    Theorem for n > 2." Proc. Cambridge Phil. Soc., 54,
    pp. 300-301.

Henrici, P., (1964). Elements of Numerical Analysis. Wiley,
    New York.

Hersh, R., and R. Griego, (1969). "Brownian Motion and Po-
    tential Theory." Scientific American, 220(3).

Hlawka, E., (1961). "Funktionen von beschränkter Variation
    in der Theorie der Gleichverteilung." Ann. Mat. Pura
    Appl., 54, pp. 325-334.

Isaacson, E., and H. Keller, (1966). _Analysis of Numerical Methods_. Wiley, New York.

Kaplan, W., (1952). _Advanced Calculus_. Addison Wesley, Reading, Massachusetts.

Loève, M., (1955). _Probability Theory_, 3rd ed. van Nostrand, Princeton, New Jersey.

Mason, S., (1976). _Approximate Multidimensional Integration Methods_. Thesis for M.S. degree, Systems and Industrial Engineering Department, University of Arizona.

McCracken, D., (1955). "The Monte Carlo Method." _Scientific American_, 192(5).

Metropolis, N., and S. Ulam, (1949). "The Monte Carlo Method." _J. American Statistical Assoc._, 44, pp. 335-341.

Muller, M., (1956). "Some Continuous Monte Carlo Methods for the Dirichlet Problem." _Ann. Math. Statist._, 27, pp. 569-589.

Shreider, Y., (1966). _The Monte Carlo Method_. Pergamon Press, New York.

Spanier, J., and E. Gelbard, (1969). _Monte Carlo Principles and Neutron Transport Problems_. Addison-Wesley Publishing Company, Reading, Massachusetts.

Stroud, A., (1967). "Integration Formulas and Orthogonal Polynomials." _SIAM J. Numer. Anal._, 4, pp. 381-389.

Stroud, A., (1969). "Integration Formulas and Orthogonal Polynomials for Two Variables." _SIAM J. Numer. Anal._, 6, pp. 222-228.

Stroud, A., (1971). Approximate Calculation of Multiple
    Integrals. Prentice Hall, Englewood Cliffs, New Jersey.

Tukey, J., (1957). "Antithesis or Regression." Proc.
    Cambridge Phil. Soc., 53, pp. 923-924.

Wilks, S., (1962). Mathematical Statistics. Wiley, New
    York.

Yakowitz, S., (1969). Mathematics of Adaptive Control
    Processes. Elsevier, New York.

Yakowitz, S., L. Duckstein, and C. Kisiel, (1974). "De-
    cision Analysis of a Hydrologic Variate." Water Re-
    sources Research, 10, pp. 695-704.

Zaremba, S., (1968). "The Mathematical Basis of Monte Carlo
    and Quasi-Monte Carlo Methods." SIAM Review, 10,
    pp. 303-314.

INDEX

INDEX

Note. Underlined page numbers denote the pages on which a literature reference of the indicated author may be found.

TITLES OF RELATED INTEREST

John Casti and Robert Kalaba
IMBEDDING METHODS IN APPLIED MATHEMATICS
Applied Mathematics and Computation Series No. 2

"The method of invariant imbedding has its origin in the theory of transmission lines and also in certain ideas introduced by Hamilton in his study of optics. Its main novelty is to treat the end points of the interval in which the problem is defined as primary variables. The book applies these ideas to difference equations, differential equations, integral equations, and variational problems and also gives some interesting physical applications. . . .It has been said that 'on the frontiers of research science, a sound number is indeed a shining goal.' The authors seem to agree and ease of numerical analysis is a distinct advantage, as well as a major objective, of their treatment.

"I particularly enjoyed the chapter on integral equations and the application to radiative transfer in the last chapter. However, others may be more interested in the novel treatment of boundary-value problems in Chapter 3 or in the formulation of variational problems in Chapter 5. . . .

"I recommend it to all concerned with the applications of mathematics or with its computational aspects."
—American Scientist

CONTENTS:

Finite Difference Equations. Initial Value Problems. Two-Point Boundary Value Problems. Fredholm Integral Equations. Variational Problems. Applications in the Physical Sciences. Appendix. Index.

1972, xiv, 306 pp.;
hardbound ISBN 0-201-00918-8
paperbound ISBN 0-201-00919-6

V. K. Murthy
THE GENERAL POINT PROCESS: Applications to
Structural Fatigue, Bioscience, and Medical Research
Applied Mathematics and Computation Series No. 5

Review: "A theory involving the distribution of events or points in
any given time frame (i.e., the general point process) is developed,
and several fields of application of the theory are considered. 'Events'
are understood to mean such occurrences as a pedestrian being knocked
down by a car, an airplane experiencing sudden accelerations due to
turbulence or maneuvering procedures, or a person suffering an acute
myocardial infarction. Estimation statistics, the cumulative point
process, the multidimensional point process, applications to the theory
of cumulative fatigue damage of structures, the superposition of arbitrary point processes, applications of the theory of the general point
process to life test data, and applications to biological and medical
problems are discussed."

—International Aerospace Abstracts

Contents: Preliminaries. Estimation. The Cumulative Point Process.
The Multidimensional Point Process. Estimation of the Variance of the
Counts Resulting from a Two-Dimensional Process. Applications to the
Theory of Cumulative Fatigue Damage of Structures. The Superposition
of Arbitrary Point Processes. Applications of the Theory of the General
Point Process to Life Test Data. Appropriate Distributions for the Inter-
Event Intervals for the Point Damage Processes and Their Estimation
Problems. A Simple Two-Parameter Weibull Model for the Inter-Event
Distribution. Applications to Biological and Medical Problems. Further
Applications to Biological and Medical Problems. Bibliography.

1974, xx, 604 pp. illus.;
hardbound ISBN 0-201-04892-2
paperbound ISBN 0-201-04893-0

Peyton Z. Peebles, Jr.
COMMUNICATION SYSTEM PRINCIPLES

This text, written primarily to introduce the student of electrical
engineering to the principles of communication systems, is an
outgrowth of courses taught by Dr. Peebles at the University of
Tennessee. The work deals mainly with theoretical principles, and
more than 20 systems—including variations—are considered. Most
of the emphasis is placed on modern pulse and digital systems.

Many examples are included as an integral part of the book, and there
are also more than 275 problems covering most of the subjects taken
up in the text. The book is written with a great deal of flexibility,
and thus the most important portions of the book may be profitably
studied by students with a variety of backgrounds.

CONTENTS:

Introduction to Book. Deterministic Signal Representations. Deterministic Signal Transfer Through Networks. Statistical Concepts and the
Description of Random Signals and Noise. Amplitude Modulation.
Angle Modulation. Pulse and Digital Modulation. Carrier Modulation
by Digital Signals. System Power Transfer and Sensitivity. Index.

1976, xx, 488 pp., illus.;
hardbound ISBN 0-201-05758-1
paperbound ISBN 0-201-05759- X

Addison-Wesley Publishing Company, Inc.
Advanced Book Program
Reading, Massachusetts